SEVERE PLASTIC DEFORMATION TECHNIQUES

Edited by **Marcello Cabibbo**

Severe Plastic Deformation Techniques
http://dx.doi.org/10.5772/66556
Edited by Marcello Cabibbo

Contributors

Marina Borodachenkova, Wei Wen, António Manuel De Bastos Pereira, Yuepeng Song, Hyoungseop Kim, Yanjin Guan, Zongshen Wang, Milos Janecek, Petr Minarik, Tomáš Krajňák, Radomír Kužel, Milan Dopita, Jakub Cizek, Ondrej Srba, Jenő Gubicza, Jie Xu, Bin Guo, Debin Shan, Abdrakhman Naizabekov, Sergey Lezhnev, Evgeniy Panin, Irina Volokitina, Jitka Stráská, Pavel Zháňal, Kristína Václavová, Josef Strasky, Petr Harcuba, Kinga Rodak

Published by InTech
Janeza Trdine 9, 51000 Rijeka, Croatia

Publishing Process Manager Dajana Pemac
Technical Editor SPi Global
Cover InTech Design team

First published July, 2017
Printed in Croatia
Legal deposit, Croatia: National and University Library in Zagreb

Additional hard copies can be obtained from orders@intechopen.com

Severe Plastic Deformation Techniques, Edited by Marcello Cabibbo
p. cm.
Print ISBN 978-953-51-3425-1
Online ISBN 978-953-51-3426-8

For EU product safety concerns:
IN TECH d.o.o., Prolaz Marije Krucifikse Kozulić 3, 51000 Rijeka, Croatia,
info@intechopen.com or visit our website at intechopen.com.

Meet the editor

Professor Marcello Cabibbo was born in Palermo in 1971 and graduated in Physics at the Alma Mater Studiorum University of Bologna in 1996. He took his PhD degree in Materials Engineering at the University of Tor Vergata in Rome (2000). He was a university researcher from 2000 to 2007, and from that year, he has been an associate professor of Metallurgy with the DIISM/Università Politecnica delle Marche. In his early years of academic career, he won several awards, of which the most relevant are Young Researcher award from Università Politecnica delle Marche (2000 and 2002) and Researcher of the Year award from Università Politecnica delle Marche (2003) and he was third at a national ranking for the Materials Science Microscopy SISM (Italian Society of Microscopy Sciences) (2004). He is currently a member of the European Microscopy Society (EMS), the Italian Society of the Microscopy Sciences (SISM), and the Italian Metallurgy Association (AIM). He is the coauthor of more than 170 journal papers, two-third of which published in peer-reviewed (ISI) international journals, and most of them as corresponding author (113 according to Scopus, ISI. WoS).

Contents

Preface

Grain size is recognized as a key microstructural factor affecting mechanical and, to some extent, physical properties of metals and metallic materials. For this reason, all the means designed to control and modify the grain size are considered a proper way to design and tailor metallic materials with desired properties. In this sense, microstructure refinement through severe plastic deformation (SPD) techniques can be considered a key method for this purpose.

A typical SPD process is currently defined as any method of metal forming under extensive hydrostatic pressure intended to impose a very high strain on a bulk solid without involving any significant change in the overall dimensions and having the ability to produce exceptional grain refinement.

What makes SPD processing techniques so popular and attractive is the possibility of using them to enhance the strength behavior of conventional metallic materials by a factor of up to eight for pure metals such as copper and by some 30–50% for alloys. Despite the impressive property improvement achievable with SPD techniques, their uptake by industry has been rather sluggish.

This book intends to give a panorama of the typical SPD techniques intended to optimize the mechanical and physical properties of metals through a significant grain size reduction process. Modeling for this purpose is also presented.

In particular, this book presents original and recent state of the arts concerning two of the major SPD techniques, that is, equal-channel angular pressing (ECAP) and high-pressure torsion (HPT). Several case studies on different SPD techniques are here presented with high-standard experimental methods and unique mechanical/microstructure results. The grain size refinement stability upon high-temperature exposition is also discussed in a specific contribution. Finally, a great attention has been driven in the modeling and simulation of the grain size reduction potentials and limits involved in different SPD techniques.

Upon my personal experience of already 20 years of research activity in the field of SPD techniques, and especially on ECAP of light alloys, I believe that the different chapters of this book describe well the state of the art and the recent development of the most important SPD techniques applied to metallic materials and alloys.

Marcello Cabibbo
DIISM/Università Politecnica delle Marche,
Ancona, Italy

Section 1

SPD: ECAP and HPT

Chapter 1

Ultrafine-Grained Materials Fabrication with High Pressure Torsion and Simulation of Plastic Deformation Inhomogeneous Characteristics

Yuepeng Song, Wenke Wang, Miaomiao Chen, Jing Guo, Lingfeng Xu, Dongsheng Gao and Hyoung Seop Kim

Additional information is available at the end of the chapter

http://dx.doi.org/10.5772/intechopen.68360

Abstract

Utilization of severe plastic deformation (SPD) methods has provided a convenient approach for producing ultrafine-grained (UFG) materials exhibiting outstanding characteristics especially mechanical properties. HPT as one of the SPD methods can lead both to smaller grains and to a higher fraction of high-angle grain boundaries, which is an especially attractive procedure by researchers. In order to understand the nonlinearities relationship between the mechanical properties and the developed strain during plastic deformation, local deformation analysis using the finite element methodwas applied for the HPT process. In this chapter, results are reported of an investigation on the deformed microstructure and mechanical properties of different materials samples during the HPT process using experiments and FEM simulations. Simulation results indicate that the disks show inhomogeneity development and distribution of strain and stress during the plastic deformation. Microstructure and hardness investigation results can give a well support to verify the rules of inhomogenous plastic deformation in the early stage of the HPT disks. Furthermore, the friction and anvil geometry play important roles in the homogeneity of the deformation. After the hollow cone high pressure torsion (HC-HPT), the thermal stability of $Zr_{64.13}Cu_{15.75}Ni_{10.12}Al_{10}$ BMGs is enhanced, while the elastic modulus of BMG will be decreased.

Keywords: High Pressure Torsion (HPT), Ultrafine-Grained (UFG) Materials, Severe Plastic Deformation (SPD), Inhomogeneous Characteristics, Finite Elements Method (FEM)

1. Introduction

In recent 20 years, the investigation on the micro-structural evolution of ultrafine-grained (UFG) materials surged tremendously due to outstanding characteristics of UFG materials, especially mechanical properties [1–4]. UFG materials are defined as materials having equiaxed microstructures with average grain sizes less than 1 μm and with a high fraction of boundaries having high angles of misorientation. These UFG structures divide into materials having submicrometer materials where the grains are within the range of 0.1–1 μm and true nanometer level materials where the grain sizes are <100 nm. As described elsewhere, many literatures reveal that the UFG microstructures may additionally contain having sizes of the order of <50 nm and these observations led to the introduction of the nanometer level materials [4–12].

Severe plastic deformation (SPD) processes have been studied extensively and used as convenient methods to manufacture ultrafine-grained, nanostructured metals, and their alloys [13–15]. Recently, several processing techniques existing metal forming processes have been designed such as continuous cyclic bending (CCB), twist extrusion (TE), equal channel angular pressing (ECAP), and so on. New techniques are often proposed, all of them rely on the idea that a high hydrostatic pressure is necessary to avoid cracking, for a review see [16–25]. Among these SPD methods, the HPT process is particularly noteworthy because it can produce finer grains, with a higher fraction of high-angle grain boundaries, than can the other SPD methods [5, 8, 26–29].

The origin of HPT processing may be traced to a classic literature, written by Bridgman and published in the Journal of Applied Physics in 1943, entitled "On Torsion Combined with Compression" [30]. This fundamental concept formed the basis of a series of experiments conducted by Professor Bridgman, the Hollis Professor of Mathematics and Natural Philosophy at Harvard University.

The constrained HPT is the main study point presently, in which there was some limited outward flow of material between the upper and lower dies. The principle of the constrained HPT process is that a sample, generally in the form of a thin disk, is subjected to high pressure between massive anvils and then processed through the application of torsional straining. One die is turned at a given rotation speed and surface frictional forces deform the sample by shearing so that deformation proceeds under a quasi-hydrostatic state. The HPT process consists of two stages based on the motion of the lower dies and the samples, as shown in **Figure 1**: first the compression stage and next the torsion stage.

During the torsion stage, the compressive pressure is generally kept constant. The high-imposed compressive hydrostatic pressure prevents any cracking of the sample inside of the die, and the low thickness to diameter ratio results in the production of a high strain during the die rotation.

Although the general principles of HPT processing were first proposed over 70 years ago, the processing has become of general scientific interest only within the last 20 years.

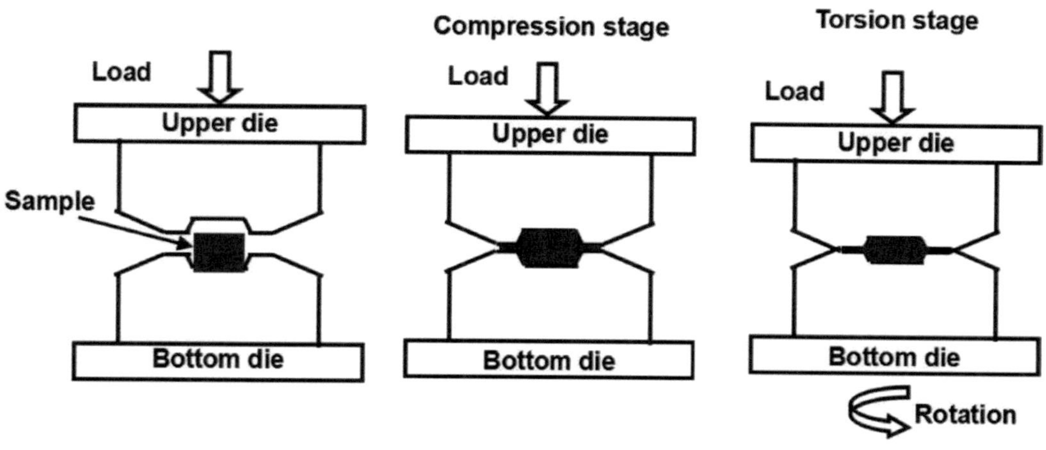

Figure 1. Schematic diagram of high-pressure torsion processing.

Moreover, it is only within the last 10 years that a number of extensive reports documenting the processing and properties of materials produced by HPT have started appearing in the scientific [5, 26, 27].

In the range of 20 years, there are many reports on preparing bulk UFG metallic materials with high pressure torsion. Kaveh Edalati form an ultra-grained structure with a grain size of ~180 nm with pure Hf by high pressure torsion under pressures of 4 and 30 GPa. In the study of I. Sabirov, Al6061-10 pct SiC and Al6061-20 pct Al_2O_3 powder metallurgy (PM) MMCs with clustered particle distribution in the as-fabricated condition are subjected to HPT at room temperature. The evolution of the microstructure during HPT is investigated. D. Gunderov from the Ufa State Aviation Technical University produced nanocrystalline $Ti_{49.4}Ni_{50.6}$ alloy in the shape of a disk 20 mm in diameter using high pressure torsion successfully. The effect of an UFG structure formed in an aluminum alloy 1570 using severe plastic deformation by HPT at room temperature and at temperature of 100 and 200? on the mechanical properties (strength and plasticity) has been investigated by M. Yu. Murashkin. Disks of $Cu_{60}Zr_{20}Ti_{20}$ composition were produced with HPT by Zs. Kovacs and the inhomogeneous microstructure of the central region consist of particles of about 50 μm and a surrounding matrix.

Presently, there are at least three aspects existing controversy in HPT field: Firstly, HPT processing consists of two stages (compression stage and torsion stage under high pressure stages, as shown in **Figure 1**). Most of the reports published focused on the last stage [32, 36–38], but there are few researchers directly on the role of compressive processing stage [39, 40]; second, because the developed strain at the center of the disk is theoretically zero and linearly increases with the distance from the center according to the characteristics of torsional strain, it is reasonable to anticipate that the microstructures produced by the HPT process will be extremely inhomogeneous. However, recently, papers demonstrated that the microstructure is reasonably homogeneous across the disks when the torsional straining continues; at last, although a lot of studies have been done on HPT [32, 36–40], most of them are for microstructure and its characterization or for processing. Because the mechanical properties of the deformed material are directly related to the effect of friction, that is, the understanding of the effect of friction is very important in HPT.

There are many reports on radial inhomogeneity in the HPT processed metallic materials [5, 31, 32]. However, a significant dichotomy is revealed by the experimental data available. Some results gave the significant variations in the values of the microhardness and microstructure across the diameters of disks processed by HPT for austenitic steel, Cu, and high-purity Ni, given the results of lower hardness values in the centers and higher values in the peripheral regions of the disks [4, 33, 34]. On the other hand, recently, results shown that, as to commercial purity Al, an Al–Mg–Sc alloy, Cu and high-purity Ni materials, when torsional straining is continued to a sufficiently high total strain, the microstructures and hardness become reasonably homogeneous across the disks [5, 34]. Considering this case, the jobs of hardness and microstructure inhomogenous distribution inspection are very important to explore the deformation mechanism during the HPT process.

Because the mechanical properties of the deformed material are directly related to the amount of plastic deformation, that is, the developed strain, understanding the phenomenon associated with strain development is very important in severe plastic deformation process. Meanwhile, in the recently three decade, computer simulation and finite elements method (FEM) have been attracted with huge interests by more and more researchers of widely fields, and they can be used to explore and gain new insights and formation mechanism study in materials preparation process.

In this chapter, results are reported of an investigation on the plastic deformation inhomogeneous characteristics of different materials samples during the HPT process using experiments and FEM simulations.

2. Experimental conditions and simulation procedures

2.1. Materials and samples

In the previous works, there are two materials disks processed by HPT: One is commercial purity copper (99.98 mass%), and the other is IF steel manufactured by the Pohang Steel Company (POSCO, Korea) with the composition of 0.0026 wt% C, 0.096 wt% Mn, 0.045 wt% Al, and 0.041 wt% Ti. The state of the two materials is listed in **Table 1**. In addition, hollow cone $Zr_{64.13}Cu_{15.75}Ni_{10.12}Al_{10}$ bulk metallic glass (BMG) was prepared by sucking into a copper mold.

For the HPT experiments, materials of copper and IF steel with six magnitudes of pressure of 1, 2, 4, 6, 8 and 10 GPa were imposed on the disks at room temperature. The applied revolutions of the bottom die are 0, 1/4, 2/4, 3/4, and 1 turns under the provided pressures. The time of applying the compression loads was set as 10 s, and the strain rate in the disk was low enough to ignore any thermal effect.

Finally, in the HPT experiments of $Zr_{64.13}Cu_{15.75}Ni_{10.12}Al_{10}$ alloy, the ingots were produced through alloying high-purity elements (minimum 99.9 wt%) in an arc furnace under an

Initial state	Copper	IF steel
Original state	Cold-drawn state	plate 12 mm in thickness size rolled from casting ingot
Homogenizing heat-treatment	600°C × 2 h, cooling in furnace	700°C × 2 h, cooling in furnace
Sizes of HPT disks	9.5 mm in diameter and 2 mm in thickness	19.5 mm in diameter and 2 mm in thickness
Grains size	20–40 μm	150–200 μm
Original hardness	~56 Hv	~80 Hv

Table 1. State of the two materials disks.

argon atmosphere. The remelted alloy ingots were suction cast into a Cu mold in order to obtain hollow cone specimens with base diameter, cone height, and wall thickness were 19.8, 14.0, and 1.2 mm, respectively. A hollow cone specimen of Zr-based BMGs is set on the concave die; then, a convex punch is inserted into the specimen; last, the preset pressure (40 tons) and rotation angle (1 reverse turn) are applied to the hollow conical specimen, as shown in **Figure 2**. The primary difference compare with the above HPT process is the sample shape, which is a hollow cone in the HC-HPT process rather than a disk in the HPT process.

2.2. Measurement devices and approaches

For the measurement of copper and IF steel, there are two inspection planes along with different direction of disks: One is radial plane (following the longitudinal direction of disks), and the other is transversal plane (following the transversal direction of disks). On transversal plane, the hardness of points on different diametrical direction is measured, whose angle between the adjacent direction is 30°. The distance between the adjacent testing points is 0.25 mm for 9.5 mm diameter and 0.5 mm for 19.5 mm diameter. As to the radial plane, the distance between the adjacent testing points is 0.1–0.3 mm at axial direction and 0.5 mm at radius direction. The schematic drawing of sampling positions and the measurement of hardness distribution of different positions are shown in **Figure 3**.

Hardness was measured using FM-700 Microhardness Tester, and the pressure loading is 100 g, continuous 10 s. All of hardness data are processed by Origin software. The color-coded contour maps and curves of hardness distribution of different samples are so obtained. Microstructures in different position of disks were observed using optical microscopy (Olympus U-TV0.5xc) and electron backscattered diffraction (EBSD).

The microstructure of $Zr_{64.13}Cu_{15.75}Ni_{10.12}Al_{10}$ alloy was determined using D/max-rB X-ray diffractometer with Cu Kα radiation, and the wavelength of the X-rays was 0.154 nm. Differential scanning calorimetry (DSC 404, Netzsch, Germany) was performed at a constant heating rate of 0.667 K/sec under a constant flow of argon. The dilatation measurements were executed using a conventional dilatometer (DIL 402C, Netzsch, Germany).

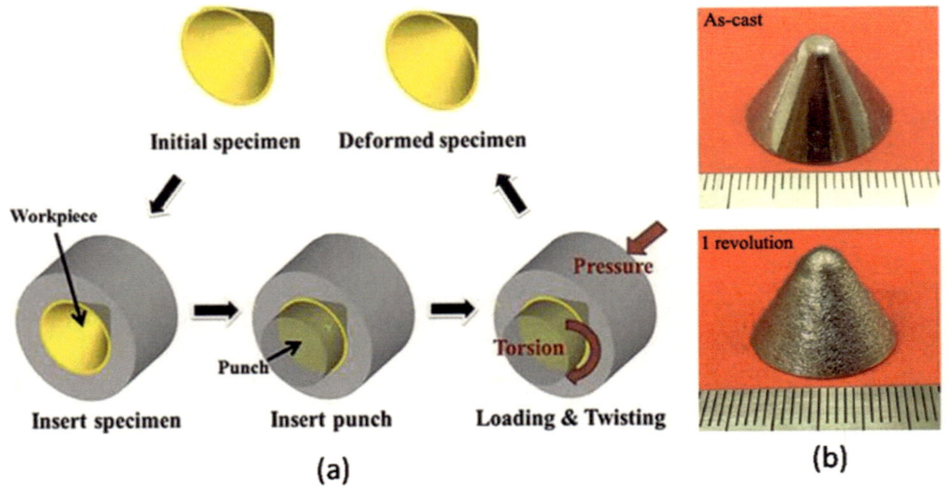

Figure 2. (a) Schematic diagram of HC-HPT procedure and (b) samples.

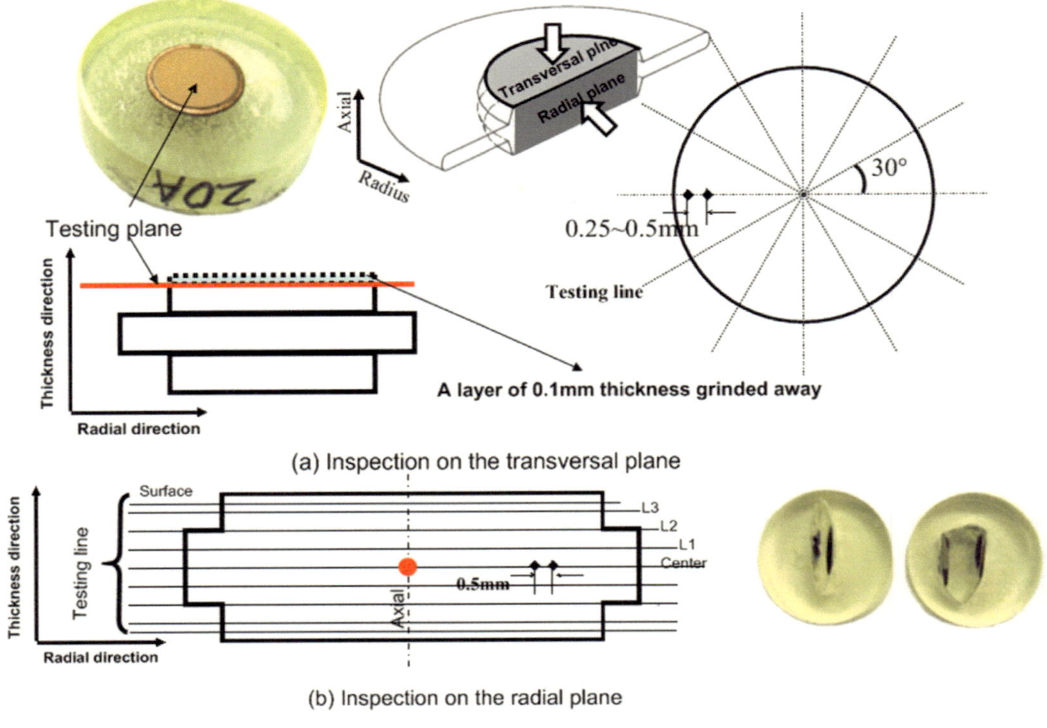

Figure 3. Schematic drawing and the measurement of hardness distribution of disks.

2.3. Simulation conditions and procedures

Plastic deformation behavior, hardness, and microstructure distribution of HPT processed disks are investigated using experimental approach and simulation approach with the finite element method (FEM). With ANSYS10.0 program simulations, the deformed microstructures and mechanical properties of copper disks in the compressive stage of HPT processing are investigated. Meanwhile, a commercial rigid-plastic finite element code (DEFORM 3D; Scientific Forming Technologies Corporation, USA) was used to simulate and understand the local plastic deformation of the IF steel disks in the torsion stage of the HPT process.

2.3.1. Simulation procedure of copper disks in the compressive stage of HPT with ANSYS

The simulation of die material is alloy steel, and the properties data come from the web site [41] and testing experiment. The simulation procedures of the copper at the compression stage of HPT are as follows:

Step 1: Build geometry. The sizes of the sample in ANSYS correspond to the experimental disk.

Step 2: Define material properties. The Young's modulus and Poisson's ratio of materials are 110 GPa, 0.343 for copper and 220 GPa, 0.32 for alloy steel, respectively [41]. The fiction coefficient between disk and anvil is 0.06.

Step 3: Define element types. For the analysis of the copper deformation, Plane 182, Target 169, and Contact 170 are selected to define the sample, the surface of the target and the surface of the contact, respectively.

Step 4: Generate mesh and create contact pair. There are about 40,000 elements meshed in the copper disks model. The meshed model is shown in **Figure 4**.

Creating contact pair using Contact Wizard with the two element types: Target 169 and Contact 170 are very important for this contact analysis.

Step 5: Apply loads. For this simulation, it needs to apply symmetry constrains on the axis of the copper sample, because the device of HPT is axis symmetrical shape.

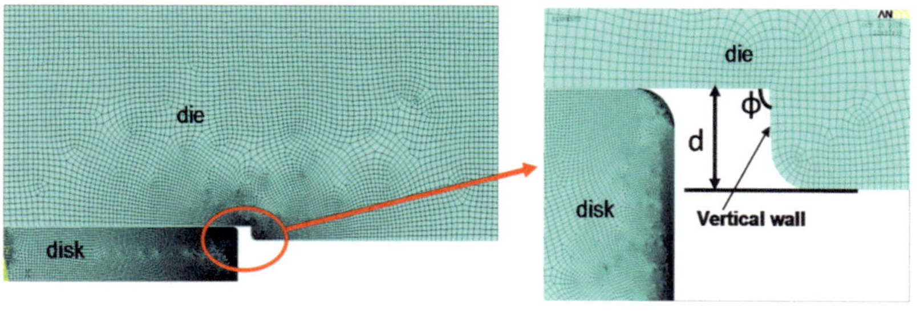

Figure 4. The meshed simulation model.

Step 6: Obtain solution and review results. Some results can be got from the general post-processor, such as the stress and the strain distribution of the compression sample, and the deformed shape.

2.3.2. Simulation procedure of IF disks of HPT with DEFORM

A local deformation analysis using the FEM should be carried out in order to better understand HPT processing and the material's response to the dies in HPT. In this work, isothermal FEM simulations of the HPT process were carried out using the commercial rigid-plastic finite element code.

In the FEM simulations, the initial dimensions of the disks were 20 mm diameter and 2.0 mm thickness. In the compression stage, the speed of the top anvil was 0.1 rad/s up to a full turn. The friction between the die and the sample was set to satisfy the sticking condition, as the roughness of the die surface was high enough to prevent slippage between the sample and the die. We used the materials parameters for the simulation from the database of DEFORM code-0.08% C carbon steel, and the Poisson's ratio of material is 0.3. HPT dies: rigid body. The number of the initial mesh in the sample was 25521, and this number of elements was enough to show the local deformation of the sample by calculation without changing the number of elements. The times of the compression stage and torsion stage were all set at 10 s.

3. Properties and microstructure inhomogeneity of different materials disks processed by high-pressure torsion

3.1. Inhomogeneous distribution of mechanical properties and microstructure in the compression stage of HPT

Since plasticity is path dependent, unlike elastic deformation, the deformation that occurs at both stage I (compression) and stage II (compression + torsion) is important for the properties and microstructures of HPT processed materials. Although many reports have been published recently on the microstructural evolution, hardness distribution in HPT processed samples, and torsional behavior [4, 7–12], all of them ignores the stage I deformation, and no studies on the properties of samples after the compression stage have been done, as far as can be determined. For example, Edalati et al. [42] investigated the microstructures and mechanical properties of pure Cu processed by HPT and proposed a unique single curve of hardness against the equivalent strain; however, they did not consider the stage I deformation and the compressive component of strains in their equivalent strain. Nowadays, more and more results indicated that the stage I deformation influences the stage II deformation [43]. Hence, explaining the HPT behavior without considering the stage I deformation is not sufficient for full understanding.

In this section, the commercial purity Copper (99.98 mass%) and IF steel are used as the study materials. For the experiment of copper, two applied pressures of 2 and 8 GPa were imposed at room temperature on the disks with the velocity range of 1/2 rpm and the time of compression load of 10 s. For the IF steel during the compression stage of HPT, the applied pressures were imposed at room temperature on the disks for 0.6, 1.25, 1.9, and 2.5 GPa, respectively.

Figure 5 shows the hardness distribution of copper disks with 2 GPa pressure given the color-coded contour maps and distribution curves. As shown, L1, L2, and L3 are the testing lines position at the distance of 0.25, 0.5, and 0.65 mm from central plane of disks' thickness direction, respectively. The hardness shows almost symmetrical distribution on the thickness direction from upper to bottom surface of compressed disks (**Figure 5a**).

Furthermore, it also indicates an inhomogeneity distribution, giving lower hardness in axial center near the surface, higher hardness in edge and the uniform hardness in radial medium, which is also clearly displayed in **Figure 5b** (L2, L3). However, a higher hardness zone exists in axial center near the central plane. Compared with the hardness of 56 Hv in the initial state, the hardness of disks remarkably increases at the compressive stage of HPT, which is different in the different position, that is, the hardness on the central plane is 106.9, 101.2, and 112.3 Hv in the center, radial medium and edge, respectively. Further investigation indicates that the hardness distribution of compressed disk with 8 GPa has a similar trend with the former results (**Figure 6**).

It is well known that the mechanical properties are mainly dependent on the microstructure condition. The detailed investigations are focused on the relationship between the microstructure and the mechanical properties, as shown in **Figure 6**. Clearly, there is a remarkable inhomogeneity distribution on the testing plane, for not only hardness but also microstructure.

In the edge zone, there is a thin layer with grains hardly change as the same as the initial state; however, a little more inward, the grains proceed large severe plastic deformation, and their boundaries become very obscure. In the center zone, some grains occur plastic deformation but others have no any change. And the grains have a uniform deformation in the radial medium zone of compressed disks. Based on the Hall-Petch relationship, the inhomogeneity distribution of microstructure can support the ones of hardness, giving lower hardness in

(a) Color-coded contour map of hardness distribution

(b) Curves of hardness distribution

Figure 5. Hardness distribution of compressed disk with 2-GPa pressure.

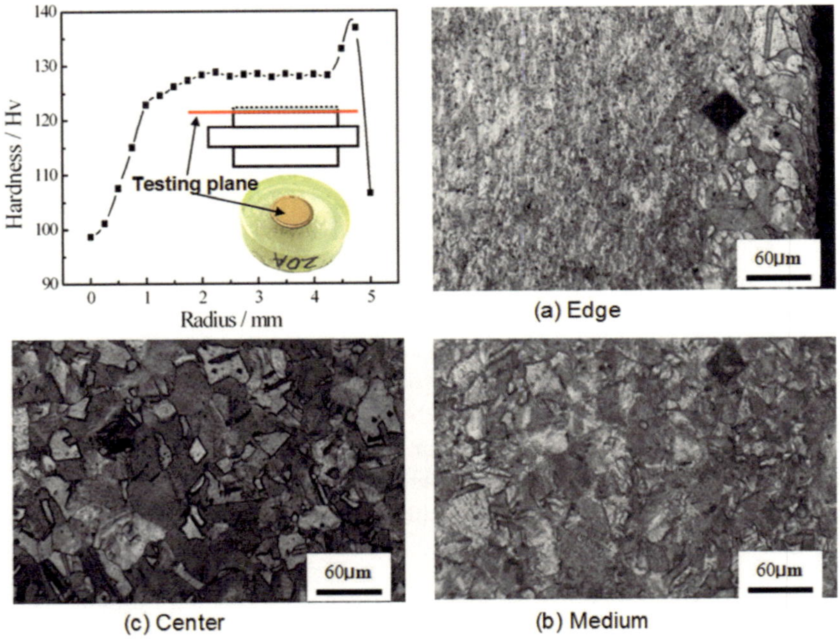

Figure 6. Hardness and microstructure distribution of compressed disk with 8-GPa pressure.

center, uniform hardness in medium, and higher hardness in edge. Of course, there is a little soft layer in the outer edge of the compressed disks.

As the former mentioned, the inhomogeneity of hardness and microstructure distribution in the compressive copper disks of HPT really existed. Of particular concern is that, as to IF steel disks, whether this inhomogeneity also existed at the compression stage during HPT. The study proceeded to investigate the inhomogeneity of hardness and microstructure on the different direction of HPT processed IF steel disks at the compressive stage.

Figure 7 displayed the hardness distribution on the transversal plane (0.1 mm distance from the surface) and the hardness variation of different position along with the compression. As the almost same distribution with the copper disks given by literature and formerly research, at the compressive stage of HPT, the hardness distribution of IF steel disks is also inhomogeneous, given high value in edge, considerable uniform in medium and low hardness in center [5, 35].

On the radial testing plane, the hardness distribution is also inhomogeneous, and the color-coded contour maps are shown in **Figure 8**. As shown, the hardness is almost symmetrical distribution from upper to bottom surface of disks for the central of thickness as the symmetry plane, which is also reported by Pippan [44]. The hardness distribution on radial testing plane is the same with that on the transversal testing plane formerly obtained, which presents high hardness in edge, uniform hardness in medium, and low value in center.

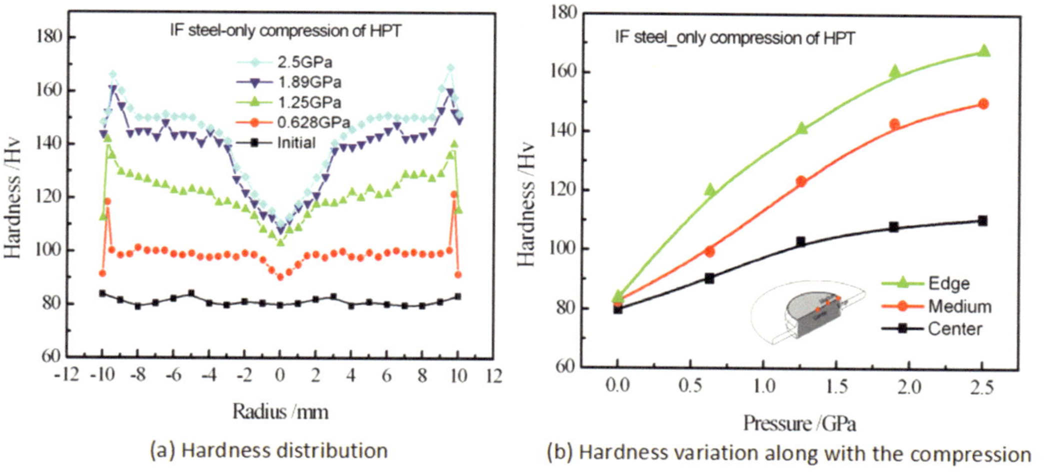

Figure 7. The hardness distribution on the transversal testing plane of different samples at the compressive stage of HPT.

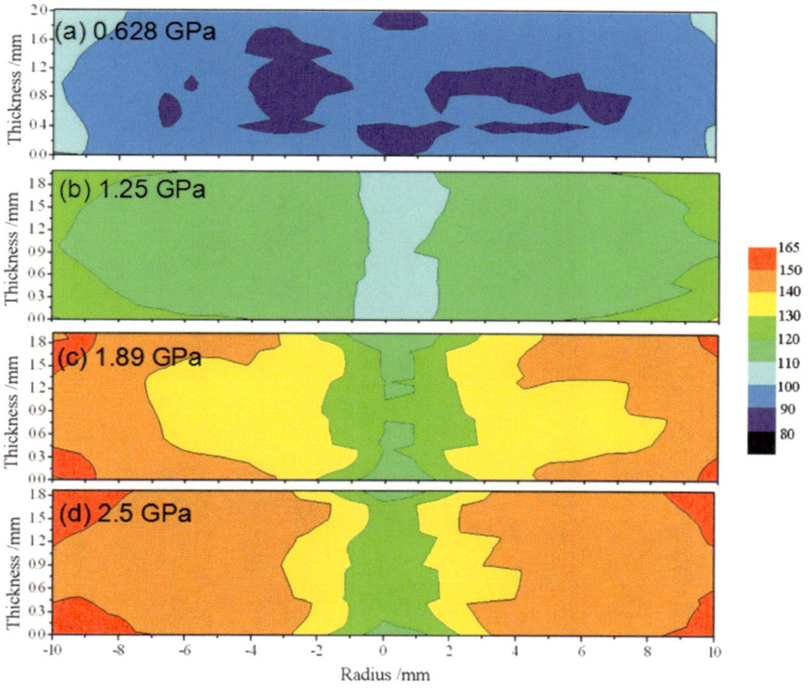

Figure 8. Color-coded contour maps of hardness distribution on the radial testing plane of compressed IF steel disks.

In addition, there exists a soften region near the surface, and higher pressure is, less area is. An important point should be paid attention to that the lower hardness areas are only in both the outedge and center near the disks' surface. However, a high value of hardness exists near the central position on the thickness direction named hardness hill from literatures [35, 36].

In order to clearly display this inhomogeneity, the edge microstructure on the different testing plane (transversal and radial) of compressed disk under 2.5 GPa pressure is shown in **Figure 9**. The same position on the two testing planes is corresponding with each other. For this disk, in the outer boundary, there exists a soften region resulting to the lower hardness (as the arrow direction). The severe deformation grains on the transversal plane are corresponding with the flow lines clearly displayed on the radial plane.

Another fact need to be paid attention is that as to the flow-line, the angle of direction between its texture and pressure is about 45°, which is the most flexible deformation direction under the shear slip for grains. Further microstructure inspection results show that the grains near the surface display more and more remarkable texture characteristic microstructure from center to edge along with the radius direction of the compressive disks.

However, the grains of the thickness central position haven't this characteristic texture microstructure. The close-up microstructure view from edge to center of compressed disks of 2.5 GPa pressure is listed in **Figure 10**. The results can also strongly support the former conclusion.

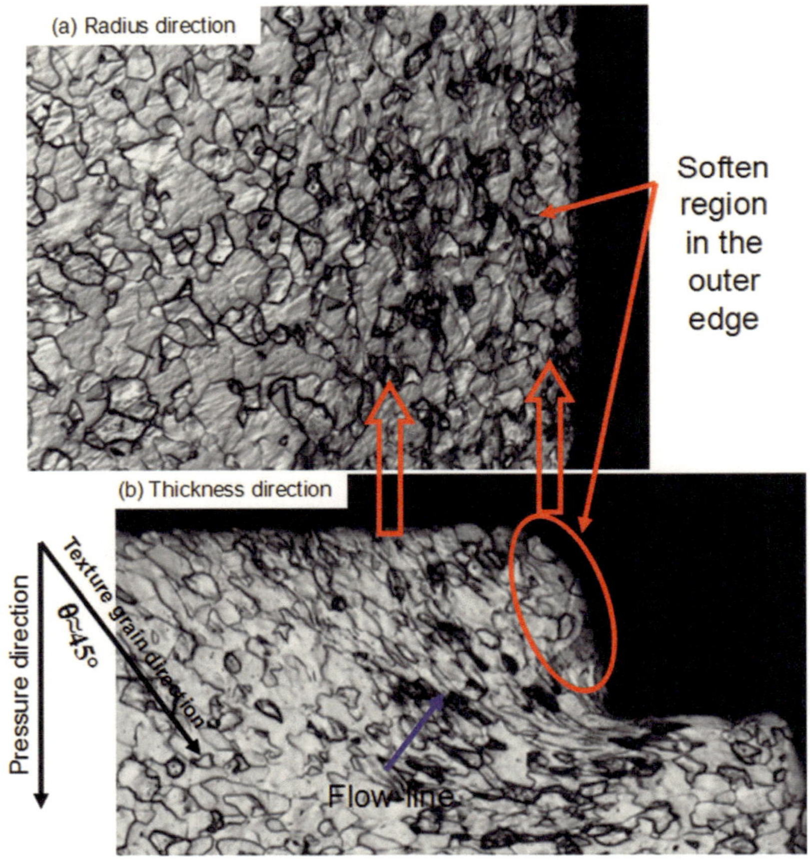

Figure 9. Edge microstructure on the different testing plane.

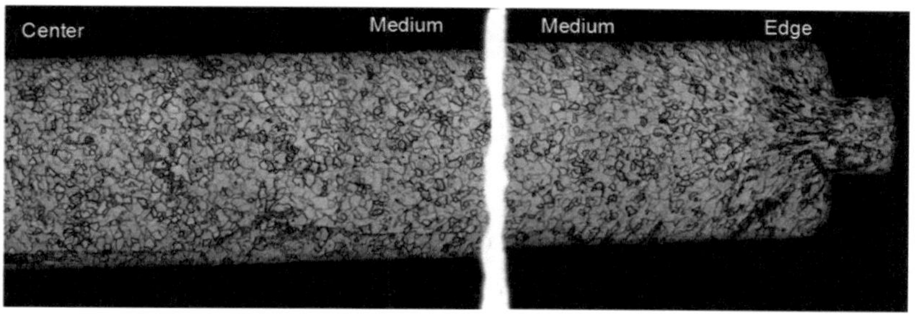

Figure 10. Close-up microstructure view from edge to center of compressed disks.

3.2. Mechanical properties and microstructure at the torsion stage of HPT

The principles of the modern HPT process have already been described extensively in the literature [5, 45, 46]. In brief, the initial coarse-grained solid or powder sample for HPT is located between two hard anvils, where it is subjected to a compressive applied pressure (P) of several GPa at room temperature (or a warm temperature), and then, a torsional strain is imposed by the rotation of the anvil. The surface frictional force generates the deformation of the disk through the torsional shear, thereby a large deformation proceeds under a quasi-hydrostatic pressure. In practice, the effective strain imposed on the sample may be defined as follows: $\varepsilon_{eq} = 2\pi NR/\sqrt{3}h$ where N is number of turns in the HPT, R is the distance from the center of the sample, and h is the sample thickness [5, 47, 48]. From the formula, following the radius direction from center to edge of HPT disks, along the R increasing, the equivalent von Mises strain is higher to higher which means the deformation of material more and more severe.

In this section, the development of deformed microstructures and mechanical properties of the IF steel disks are presented at the early torsion stage of the HPT process using experiments approaches. The applied pressure and degree of revolutions during the torsion stage were 2.5 GPa and 0, 1/4, 2/4, 3/4, and 1 turns, respectively. **Figure 11** presents the hardness distributions on the radial-axial plane of the HPT processed IF steel disks.

Exhibiting the same trend as in the literature, the Figure clearly indicates lower hardness values in the center and higher values at the edges: after 1 turn, the hardness values in the center, middle, and edge were 140, 160, and 375 HV, respectively. Compared with the disks after torsion, in the center, the soft region penetrates through the upper and bottom surfaces of the compression-only disk (0 turns), and it disappeared after 1 turn. Furthermore, the soft region shrunk as the degree of revolutions increased. The large deformation and high hardness (240 HV) proceeded at a distance of 6 mm from the center of the disks (1/4 turn) and at a distance of 4 mm from the center of the disks (1 turn). That is, a severe deformation will proceed gradually from the edge to center along with increases in the degree of revolutions at the early torsion stage of the HPT process.

A layer that is 0.2 mm from the surface of the HPT-processed disks was ground and polished. The hardness on the testing plane along the radial direction is shown in **Figure 12**.

Figure 13 shows the hardness on the testing plane along the radial direction. As shown, the hardness trend on the transversal direction at the early torsion stage in the HPT process is

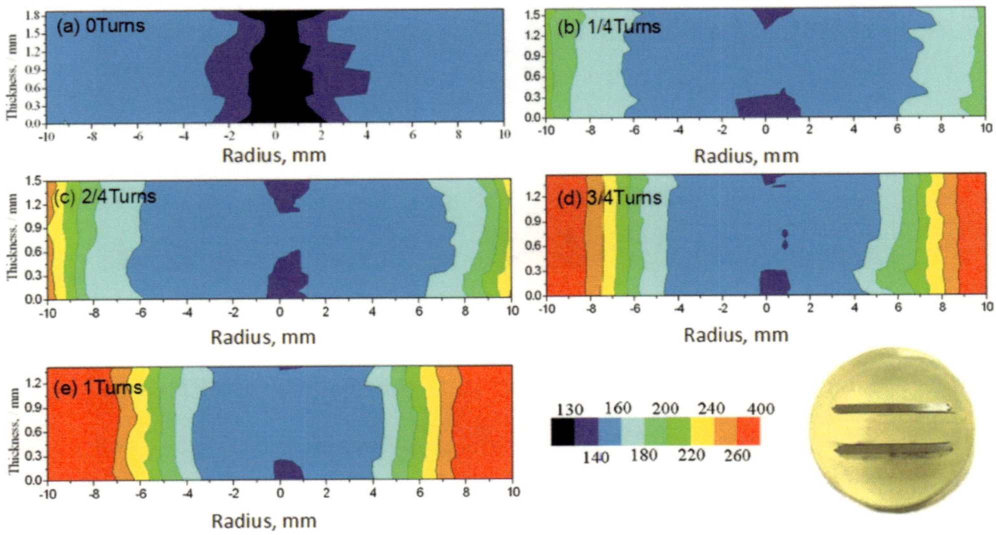

Figure 11. Hardness distribution on the radial-axial plane.

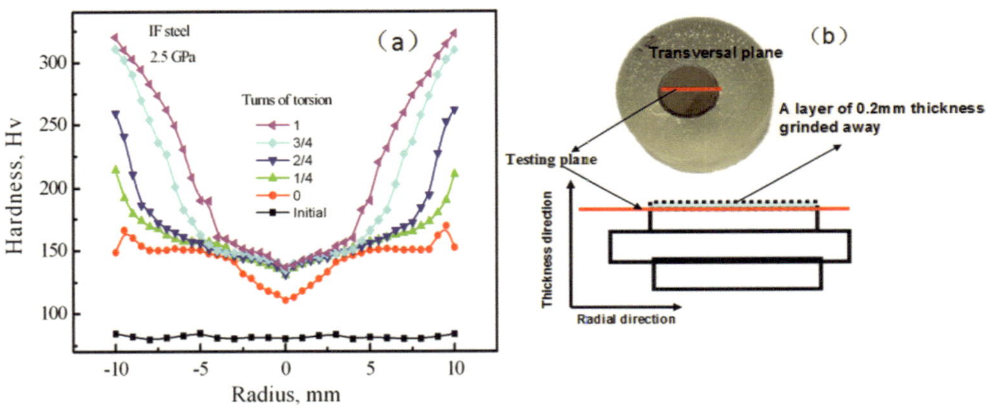

Figure 12. Hardness variations from the center to the edge (a) and the measuring plane (b).

the same as those along the radial direction. Similar to the compression stage, the different mechanical properties at the torsion stage could be explained by the microstructure state, as shown in **Figure 14**. Without torsional straining (0 turns), the grain boundaries are clearly observed, and the grains are insufficiently equiaxed. In contrast, after an additional revolution in the same torsional direction to give 1 turn, the grain boundaries become obscure, and the grains are reasonably equiaxed.

The images in **Figure 14e** show inverse pole figures (IPF) with the boundary map having the information on the orientations of microstructure and the boundaries rotation angle. The

http://dx.doi.org/10.5772/intechopen.68360

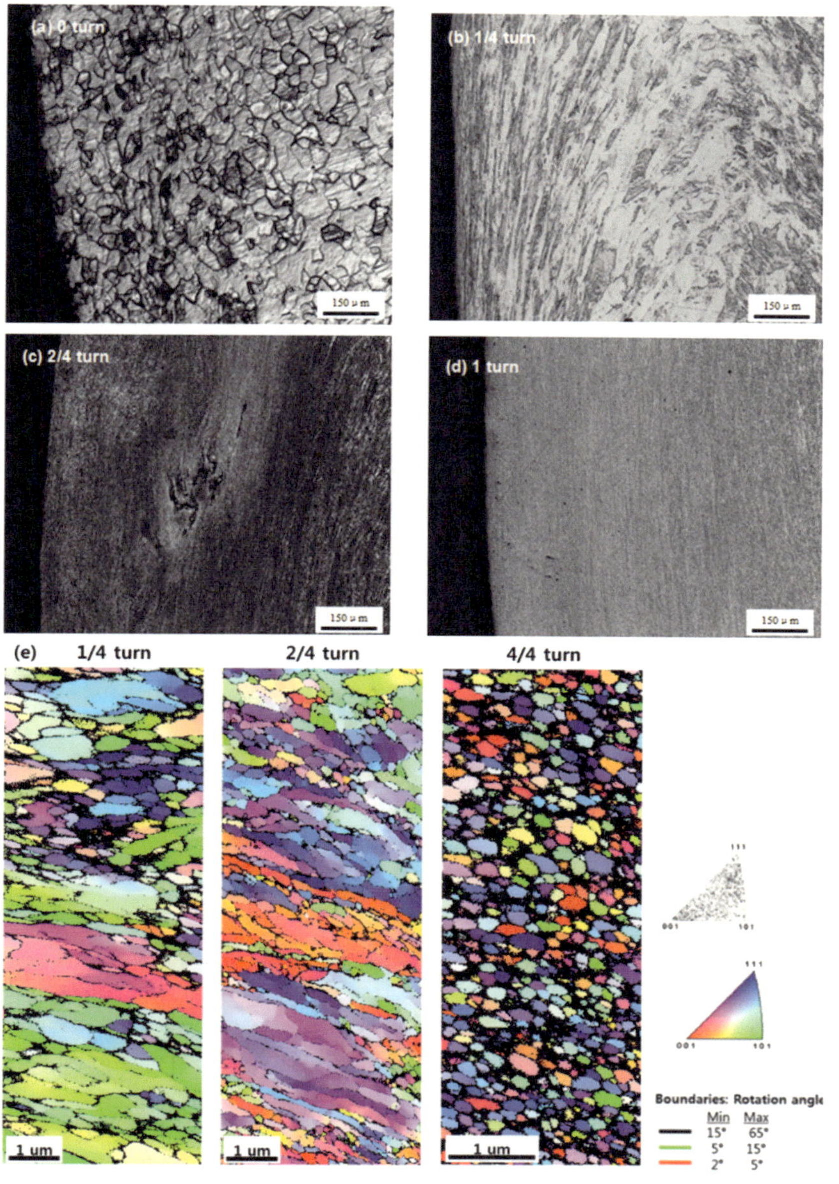

Figure 13. Microstructure at the edges of 0, 1/4, 2/4, and 1 turn HPT-processed disks using optical microscopy (a–d) and EBSD (e).

boundary map indicated high-angle grain boundaries (HAGBs: 15° <, black line) and low-angle grain boundaries (LAGBs: 2–5°, red line; 5–15°, green line). According to the EBSD results, the average grain sizes for 1/4, 2/4, and4/4 turns are 2516, 1940, and 0.308 lm, respectively. The average grain size decreases with increasing the rotation angle. After 1/4 turn, a lot of LAGBs were generated in the large grains and formed subgrains. Numerous LAGBs (green

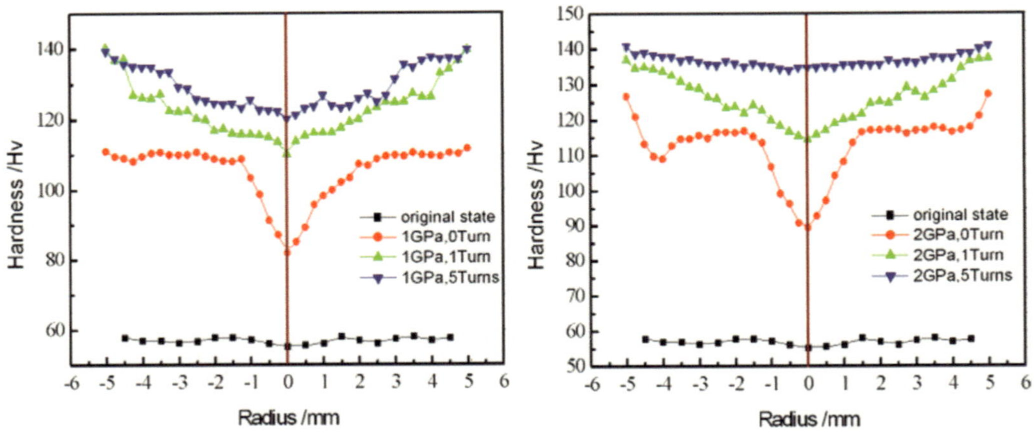

Figure 14. Hardness distribution on RP1 testing plane of different disks.

and red lines) were found within large grains, and the colors of the grains in the IPF map slightly change in the large grains. Moreover, on the top part of the image, the HAGBs are formed, and the grains are refined by recrystallization. From 2/4 turns, the fractions of HAGBs and equiaxed refined grains increased, and fine grains were located between the large grains with slightly different orientations. Finally, the average grain size with 0.308 lm was achieved after 4/4 turn with random orientation.

In the compression and the early torsion stage of HPT, the materials exist serious inhomogeneity not only in the microstructure but also in the mechanical properties. However, this is unfavorable to accommodate the following plastic strain and then highly limit the applications of HPT. Whether this inhomogeneity can be improved in the following deformation stage is a key factor for industrialized applications.

Hardness distributions on the surface plane form center to edge in 0, 1, and 5 turns HPT processed copper disks under different applied pressure are shown in **Figure 14**. The Figure clearly indicates that the hardness on the surface plane increases with increasing the degree of revolutions. However, the hardness in the center and middle zone of the HPT-processed disks varies sharply with increasing the degree of revolutions in comparison with that in the edge zone: the hardness distribution is homogeneous after several revolutions, particularly under the applied pressure of 2 GPa. That is to say, compared with the torsion stage, the hardness on the surface of the disks exhibits more inhomogeneity along the radial direction in the compression stage.

Moreover, **Figure 14** also presents that torsion can result in not only increased hardness but also uniform hardness distribution. Along the radial direction, the more the degree of revolution is, the more homogeneous the hardness distribution on the RP1 plane will be. That is, the hardness distribution becomes homogeneous with increasing the degree of revolutions.

The severe plastic deformation of copper, IF steel disks of HPT through hardness, and microstructure distribution on the testing plane of the different direction is presented. There exists serious hardness inhomogeneity on the HPT-processed disks at the compression and the early torsion stage, showing higher hardness in edge, lower hardness in center, and considerably uniform hardness in radial medium of disks. However, according to the experiment of copper at torsion stage, the above inhomogeneity in mechanical properties is hopeful to be improved by the subsequent severe torsion deformation.

4. Experimental and finite element analysis of plastic deformation inhomogeneous characteristics of HPT disks

As is known, the mechanical properties of the deformed material are related to the amount of plastic deformation, that is, the developed strain and stress during the HPT processing. The hardness and microstructure distribution associated with the strain and stress development is very important in SPD process. Thus, for systematic analysis of deformation behavior of materials, a numerical approach is useful.

In this section, plastic deformation behavior, hardness, and microstructure distribution of HPT processed disks are investigated using experimental approach and simulation approach with the finite element method (FEM). With ANSYS10.0 program simulations, the deformed microstructures and mechanical properties of copper disks in the compressive stage of HPT processing are investigated. Meanwhile, DEFORM 3D was used to simulate and understand the local plastic deformation of the IF steel disks in the torsion stage of the HPT process.

4.1. Deforming simulation results and verifications in the HPT process

The mechanical properties of the deformed material are attribution to the amount of plastic deformation, that is, the development and distribution of strain and stress of disks during the compressive processing [49, 50]. The key factors of inhomogeneity distribution of hardness and microstructure are attributed to the inhomogeneity deformation of copper disks. That is to say, the inhomogeneous distribution of strain and stress leads to microstructure inhomogeneity. As an example of disks with 8-GPa pressure, the strain distribution simulation and the microstructure of different position are shown in **Figure 15**.

The simulation results show that there indeed exists the inhomogenous distribution of strain and stress of compressed disks. In the outer edge region, lower strain leads the grains to hardly deformation as the arrow directing in **Figures 15a** and **16a**, which corresponds to the lower hardness. Because of huge friction between disk and the vertical wall of anvil, the deformation is restrained, and this lower strain zone proceeds. However, a little more inward, severe plastic deformation occurs, and larger strain is clearly displayed in **Figure 15a** with particular flow-line microstructure, similar with the results in **Figure 9**. Moreover, the angle 45° between the slip lines and the pressure direction can be shown in the strain distribution simulation.

In the center of disks, near the surface, there is a low strain zone, and the microstructure shows that some grains deformed but others have no change (**Figure 15c**). These results can explain

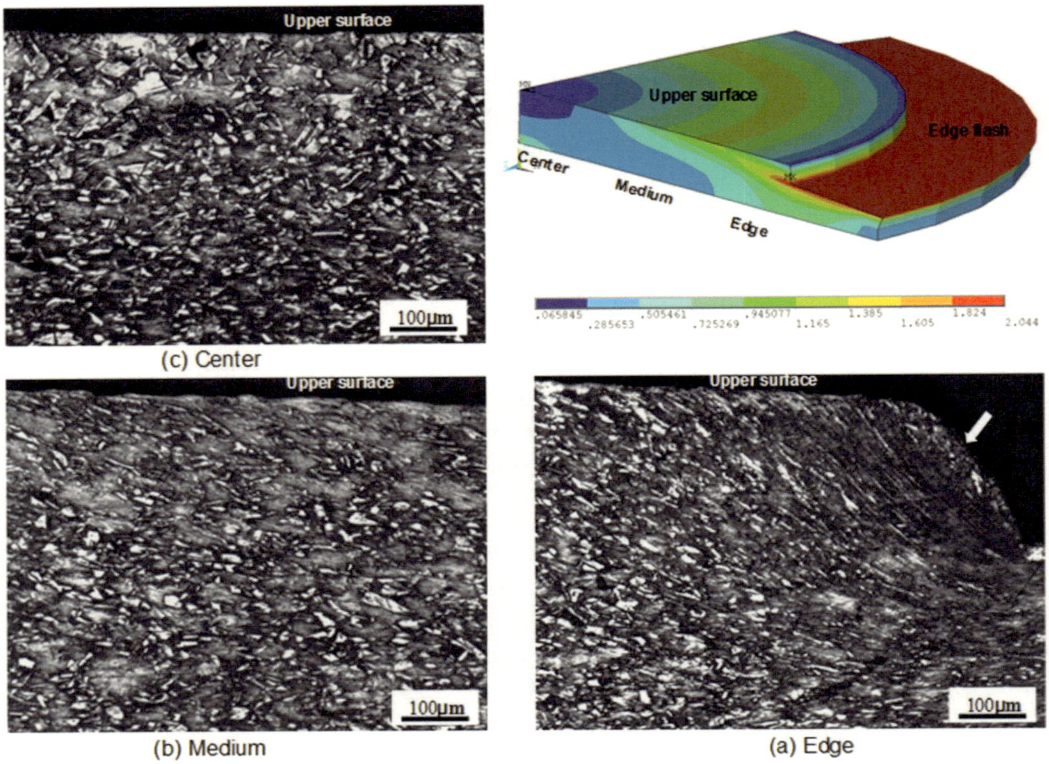

Figure 15. Strain distribution simulation and microstructure in different position of disk with 8 GPa pressure.

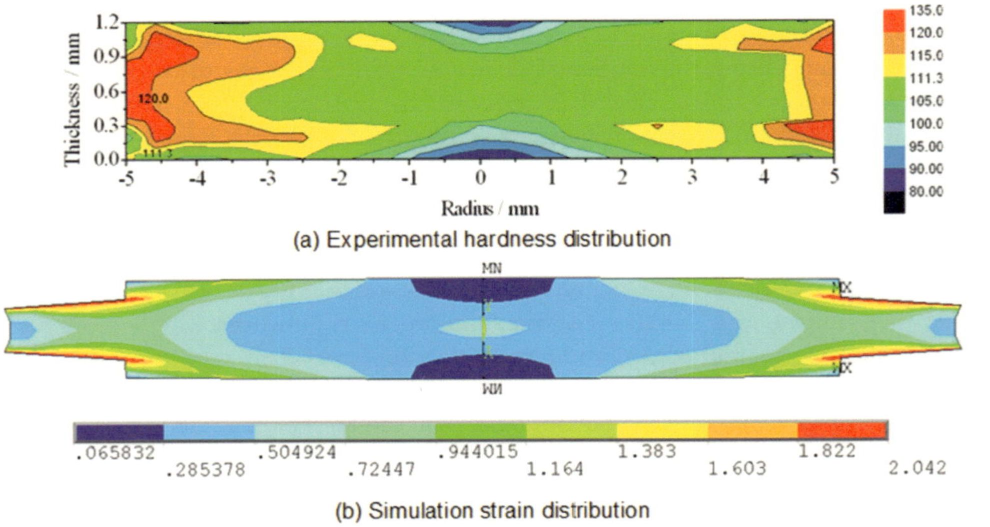

Figure 16. Hardness and strain simulation distribution of disk with 8 GPa pressure.

the reasons of the lower hardness existing in this region shown in **Figure 5**. On the other hand, near the central plane of thickness direction of disks, the strain remarkably increased with the consequence of grains deformation and high hardness.

Figure 16a is the experimental hardness distribution of disk in the axial direction and the simulation results of strain distribution of compressive disk with 8-GPa pressure.

As shown, simulation results can also verify the rules of hardness distribution on the compressive disk plane. For example, the higher hardness zone in the edge relates to its severe plastic deformation causing ultrafine grains. On the other hand, in the center, low deformation near the surface leads to low hardness corresponding with its hardly unchanged grains of the microstructure characteristic.

Subsequently, DEFORM 3D was used to simulate and understand the local plastic deformation of the IF steel disks in the torsion stage of the HPT process. In the simulation, the applied revolutions of the bottom die are 0, 1/4, 2/4, 3/4 and 1 turns under the provided pressures of 2.5 GPa, which the rotation rate of 1.256 rpm, and the coefficient of friction between the dies and the sample was assumed to be 0.12.

Figure 17 shows the integrated results including the relationships between the revolutions and hardness, the effective strain in different positions of the HPT-processed disks and the microstructure of 1/4 turn of the HPT-processed disks.

From **Figure 17b**, the simulation results indicate that the effective strain in the edge position varies sharply with increases in the revolutions, but the trend of the effective strain in the medium and center positions varies quite slowly. The strain distribution status leads to the microstructure results in **Figure 17c**, which the grains are obscure at the edge, and clear grains are observed in the central region. As a consequence, the hardness variation in **Figure 17a** is consistent with the trend of the effective strain, that is, the hardness varies sharply with the increasing degree of revolutions for the edge position of the HPT-processed disks but quite slowly for the medium and center positions.

In the simulation of HPT, disks occur the inhomogeneity development and distribution of strain and stress during the deformation, given the large severe plastic deformation in edge, lower deformation in center near the surface and considerable uniform deformation in radial medium of disk. The strain inhomogeneity determines the characteristic of microstructure and grains deformation even following the appearance of hardness distribution.

4.2. Effect of friction on HPT processing via finite element analysis

The HPT process involves changing the shape of the sample by forcing it to flow through a system, which requires tight contact between the die and sample. As a result of this contact, tangential frictional forces are generated at the interface of the die/sample to resist this relative movement. It is known that frictional conditions at the interface of the die/sample can affect the metal flow, finial properties of the sample, total deformation load, and premature die wear. The effect of friction between the sample and the dies is complex and results in the appearance of surface shear, particularly in HPT. Thus, friction is considered to be a major variable in metal forming processes where the sample undergoes large plastic deformations [40].

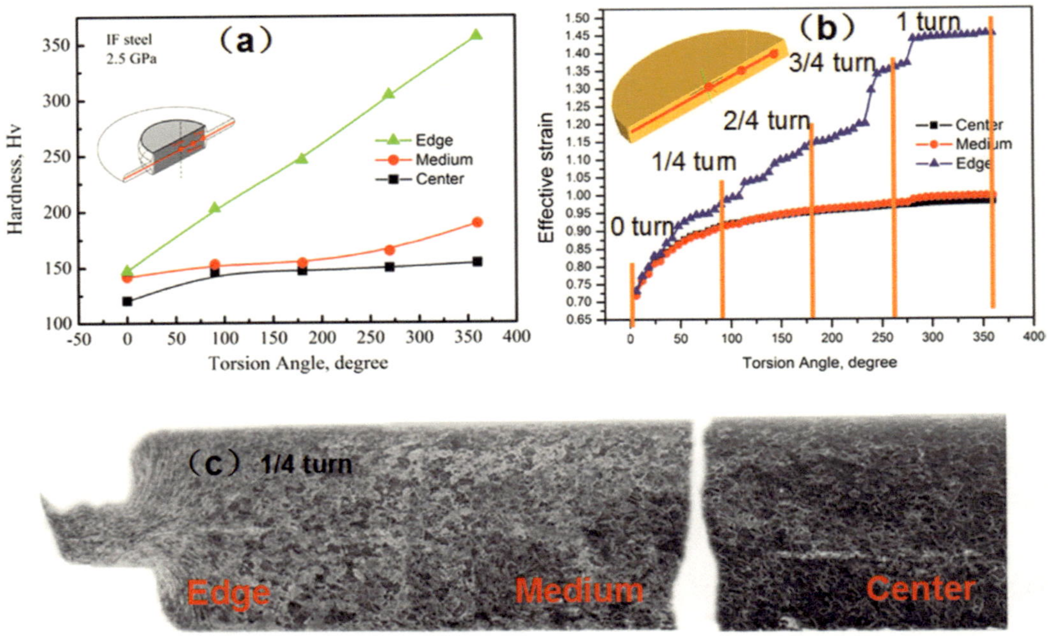

Figure 17. (a) Hardness and (b) effective strain versus rotation angle in different positions of the HPT disks, and (c) microstructure of the 1/4 turn HPT disk.

There have been preliminary investigations on the effect of friction between the anvils and workpiece on plastic deformation during the compression stage of HPT [5, 37]. The results indicated that the effective strain remarkably increased with the number of revolutions under torsion, compared to the strain in the compression stage. Although there was little variation in the central region under different friction coefficients, the strain increased significantly with distance from the center, due to frictional shear stress. Furthermore, the friction force influenced the effective strain more remarkably in the central and edge regions of the compressed disks than in the middle region. **Figure 18** shows the cross-sectional planes with effective strain distributions after compression of a copper disk under different conditions: friction coefficients of 0.1, 0.5, 1.0, and 3.0; applied pressure of 2 GPa; and wall angle of 120°.

The distribution of effective strain in the compressed copper disk is more heterogeneous as the friction coefficient increases. The radial heterogeneity of the effective strain on the plane is clearly displayed. The effective strain is lower in the center and higher at the edge of the compressed disks, and the effective strain distributions are more and more heterogeneous from center to edge. For example, with the friction coefficient of 3.0, as shown in **Figure 18d**, the variation of effective strain along the radial direction was from 0.397 in the center to 4.484 at the edge, while with friction coefficient of 0.1, the effective strain varied from 0.068 to 3.155, as shown in **Figure 18a**. This situation of heterogeneous plastic deformation was also reported in the forging process due to friction [51–55]. The results indicate that the fresh area of the contact surface, between the dies and workpiece, increased with increasing friction. Meanwhile,

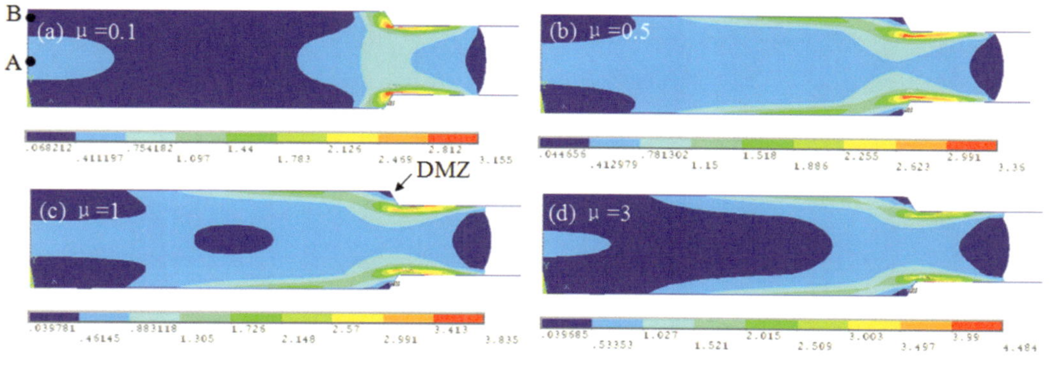

Figure 18. Effective strain distributions of the half cross-sectional planes at four friction coefficients of (a) 0.1, (b) 0.5, (c) 1.0, and (d) 3.0 under 2 GPa and 120° wall angle.

the distribution of the effective strain and hardness became more and more heterogeneous as the friction coefficient increased along the radial and axial directions, which is in good agreement with our results.

More attention should be paid to the strain distribution in the central plane of the thickness direction. The results exhibit a remarkable difference between the top and bottom planes in that the effective strain values at the center are higher than those in other areas (i.e., in the direction indicated by an arrow in **Figure 18a**. Here, an effective strain of 0.574occurred in position A (central plane), while at position B, near the surface of the upper plane, the effective strain was 0.112. Correspondingly, the hardness values of the two positions were 103.1 HV at A and 80.5 HV at B. It should be noted that the mechanical properties of the deformed material are attributable to the amount of plastic deformation (i.e., the developments and distributions of strain and stress in the workpiece during the compression process).

The hardness distribution of the experimentally compressed disks was reflected in the strain distribution of the simulations. **Figure 19** indicates the trend in variation, along the radial direction. **Figure 19** also provides a comparison between experimental hardness based on the average of four groups of experimental data, and simulation results of the effective strain distribution in the compressed copper disk, under the conditions ofa friction coefficient of 0.1, applied pressure of 2 GPa, and wall angle of 120°. The same distribution trend was indicated in both the experimental and simulated results. Hence, the reliability of this computer simulation is verified.

Figure 20 shows the evolution in effective strain at the selected point in the middle of the HPT sample for the friction coefficients of 0.5, 0.9, 1.0, and 1.5. The pressure was fixed at 10 GPa, and the number of turns was 1. Several important conclusions can be drawn from inspection of **Figure 20**. First, the effective strain values are almost the same in the compression stage at a constant of 0.4–0.6, which means that the friction had similar effect on the evolution of the effective stain on the contact surface of the HPT samples. By contrast, the effective strain is expected to increase due to the increase of the friction coefficient between the samples and the dies in the torsion stage. That is to say, the friction plays more important role on the evolution of the effective strain in the torsion stage.

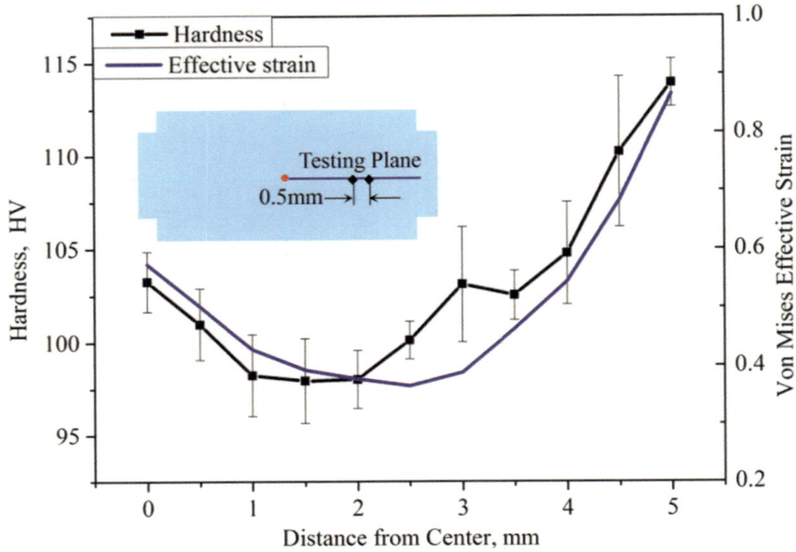

Figure 19. Path plots of the effective strain and hardness along the radial direction under 2 GPa, 120° wall angle and the friction coefficient 0.1.

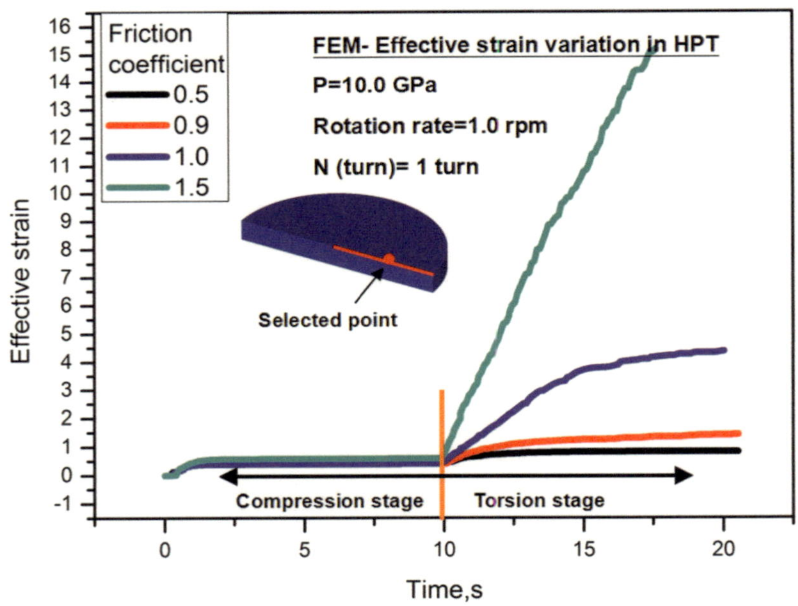

Figure 20. Simulated evolution in strain with the variation of the friction coefficient at the selected point in the medium of the HPT samples.

Another important factor is that the effective strain will reach quasi-saturation with the saturated effective strain values of 0.82, 1.33, and 4.17 when the friction coefficients are 0.5, 0.9, and 1.0, respectively. Meanwhile, the times of reaching strain saturation are also different with the variation of the friction coefficient, which are 12.1, 13.4, and 19.0 s, when the friction coefficients are 0.5, 0.9, and 1.0, respectively. Since the friction drives the surface of the sample to rotate, the effective strain remarkably increases with an increasing number of the revolutions in the torsion stage compared to strains in the compression stage. These results suggest that the friction between the sample and the dies directly affects the planes of principal stress and therefore is a major factor in the HPT process, in which the samples undergo large plastic deformation.

The simulations were performed for the friction coefficients of 0.5, 0.7, 0.9, 1.0, 1.5, and 2.0 to investigate the strain distribution on the contact surface of the HPT samples with different friction coefficients, as shown in **Figure 21a**.

Although the effective strain values in the central region were similar under different friction coefficients, the variations of effective strain according to the distance from the center had different under low (<0.9) and high (>1.0) friction coefficients. The effective strain values changed little along the distance from the center when the friction coefficients were 0.5, 0.7, and 0.9. However, the strain values increased significantly with an increasing distance from the center when the friction coefficients were 1, 1.5, and 2. The friction force affected the effective strain more in middle and edge regions than in the central region. In the middle and the edge regions, the friction shear stress due to the higher friction coefficient was high enough to achieve a sticking condition between the surfaces of the dies and the samples. **Figure 21** clearly indicates lower values of the effective strain in the central region and high values in the edge region, particularly at higher friction coefficients.

The variation of effective strain in the different position of workpiece with increasing friction coefficient is further investigated as shown in **Figure 21b**. It clearly indicates that there exist two key points of increasing friction coefficient as 0.9 and 1.5. Under the scope of friction coefficient (from 0.9 to 1.5), the effective strain sharply increases, particularly in the middle and edge area. However, it is constant variation beyond the scope (<0.9 or >1.5). That is to say, there is a friction coefficient value range which the effective strain increases remarkable sharply.

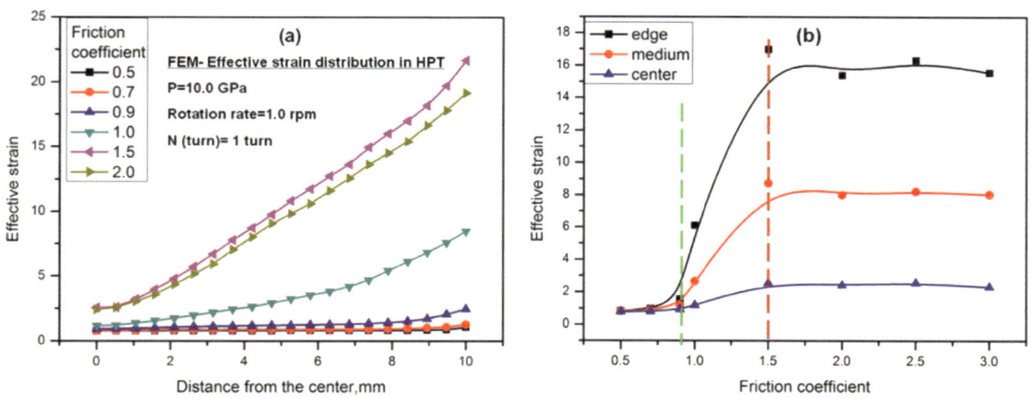

Figure 21. Simulated effective strain distribution on the contact surface of the HPT samples along with the different friction coefficient.

Compared to the compression stage, friction played an important role in the evolution of the effective strain in the torsion stage and the friction force influenced the effective strain more in the middle and edge regions than that in the central region. A high friction coefficient was enough to achieve a sticking condition between the surface of the die and the sample in medium and edge regions. Meanwhile, there is a friction coefficient value range which the effective strain increases remarkable sharply. The analysis by the finite element method for the HPT process is useful if the material parameters are incorporated. Further local and nonlocal investigations are necessary.

4.3. Effect of friction and anvil cavity structure on the dead metal zone of compressed HPT disks

The dead metal zone (DMZ) on the disks, distinct evidence of inhomogenous plastic deformation characteristics in the HPT process, was first reported by Lee et al. and verified after consideration of simulation results and inspections of microstructures from the literature [56]. In the HPT process, plastic deformation increases from the center to the edge in the radial direction of the workpiece. A sticking condition is maintained between the disk and the anvils when the traction is great enough to resist a high friction force. While there is an almost negligible strain rate and strain in the corner region, under high pressure and friction, a stagnant zone is generated due to the vertical wall constraint.

In the investigation, three factors: friction coefficient (μ), depth of the cavity on an anvil (d), and wall angle of cavity (Φ) were analyzed in the compression stage of the HPT process. The anvil structure is shown in **Figure 4**.

In the present work, the simulation results also indicated that an obvious DMZ appeared with an increase in friction coefficient (as arrow direction in **Figure 18**, and the microstructure distribution of DMZ is shown in **Figure 22** with a wall angle of 120°. There is almost no DMZ when the friction coefficient is low ($\mu = 0.1$; as shown in **Figure 18a**). Therefore, it is clear that friction remarkably influences initiation of the DMZ during the plastic deformation process in the compression stage. That is, a DMZ occurs at the corner edge of the disk under a high friction coefficient, not only in the torsion stage, but also in the compression stage of the HPT process.

Another important factor that influences initiation of the DMZ is the geometry of the anvil cavity, especially the depth of the cavity on the anvil. **Figure 23** displays the effective strain distributions on the cross-sectional planes of the compressed disks at different depths of the cavity, under conditions of 2 GPa and a friction coefficient of 0.6.

According to the effective strain distribution of the compressed disk, the degree of heterogeneity and the value of the effective strain increase as the depth of the cavity increases, along the thickness direction. **Figure 24** shows lower values of effective strain in the centers and higher values at the edges of the samples. In addition, the length of flash and the area of DMZ increase with the depth of the cavity.

Three lines of the compressed disk (**Figure 24**) were investigated in relation to DMZ, and the lines L1-L3 were effective-strain path-plot lines equally spaced from each other. The results indicated that the DMZ occurred at the surface corner of the disk and that the compressed

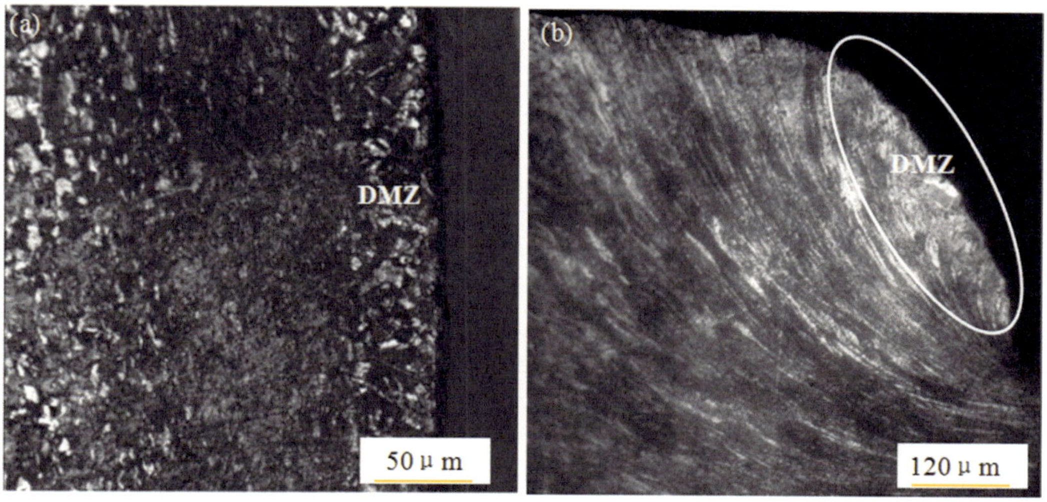

Figure 22. The microstructure distribution of DMZ on the radial direction (a) and thickness direction (b) of compressed copper disk.

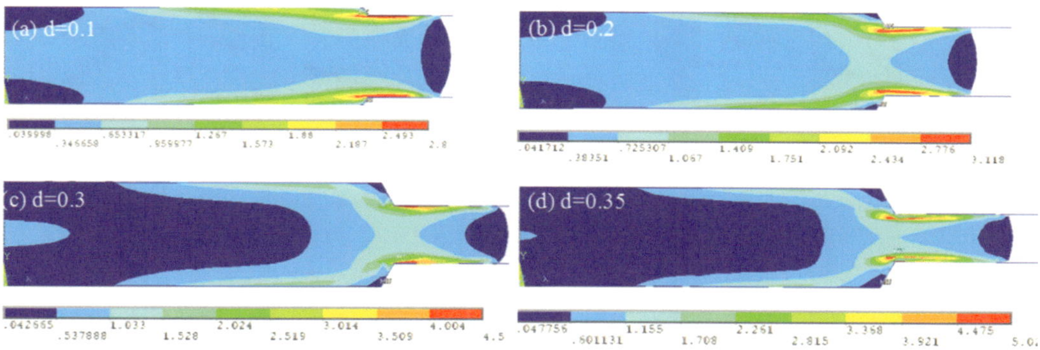

Figure 23. Effective strain distributions of the half cross-sectional planes at different depths of the cavity.

disk becomes more heterogeneous from the surface to the central plane. However, the degree of deformation in the surface region of the disk decreased with increasing depth of the cavity.

Figure 25 shows the effective strain distributions of the half cross-sectional planes along the thickness direction of a compressed copper disk, under the conditions of friction coefficient 0.1, depth 0.2 mm, and applied pressure of 2 GPa without revolution. The results can be described based on two types of HPT, unconstrained ($\Phi = 180°$) and constrained ($\Phi<180°$) HPT. The plastic deformation in the HPT process increased from the surface plane to the central plane of the disk. The hardness of the central plane was also higher, while the high hardness in the thick central plane as shown in **Figure 25** is called a hardness hill in the literature [32, 39]. The results indicate a variation in the hardness hill as the wall angle of the cavity is increased. However, there was almost no hardness hill at a low angle ($\Phi = 90°$) as shown in **Figure 25a**. That is, the wall angle of the cavity can remarkably influence the hardness hill during the plastic deformation process of the compression stage.

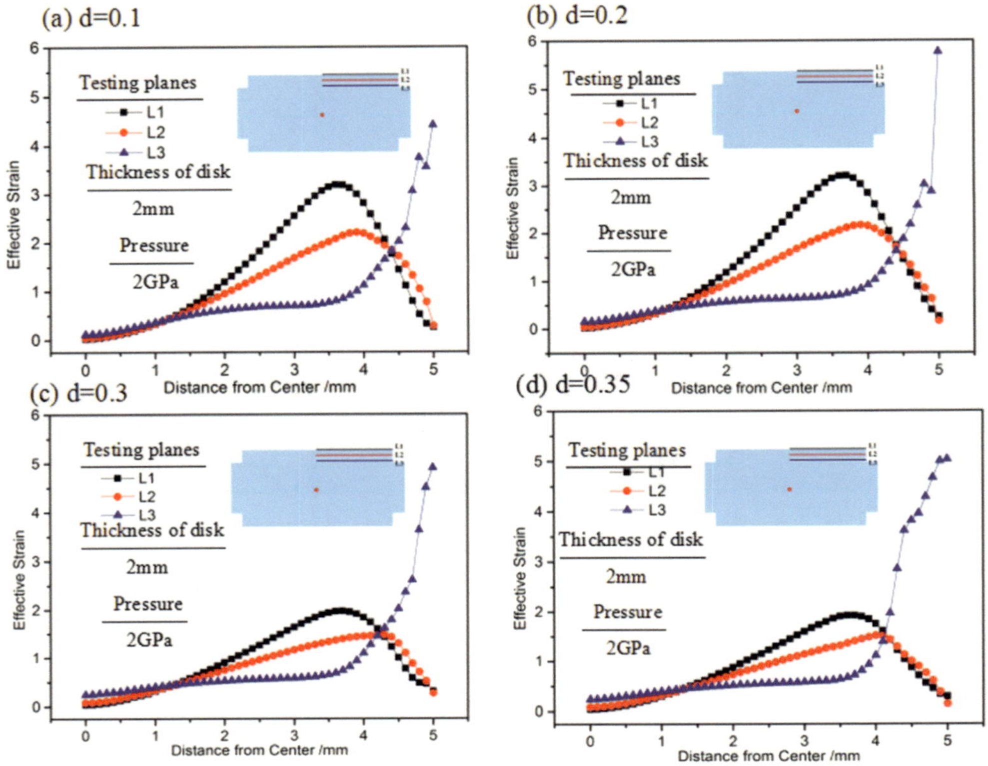

Figure 24. Path plots of the effective strain on the three lines in the disk at different depths of the cavity.

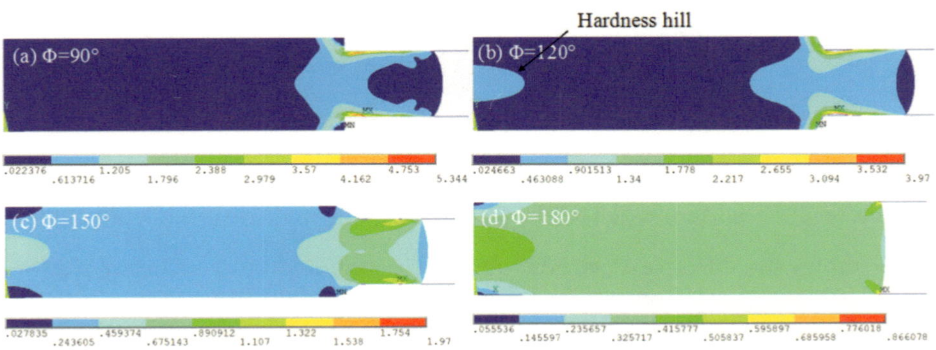

Figure 25. Effective strain distributions of the half cross-sectional planes at four wall angles of (a) 90°, (b) 120°, (c) 150°, and (d) 180° under 2 GPa and the friction coefficient 0.1.

In addition, the area of the DMZ decreased with increasing wall angle due to the vertical wall constraint under high pressure, as clearly seen in **Figure 25**, and it dropped to zero when the value of the wall angle increased to 180°, which means there is no DMZ on the disk in the unconstrained HPT processing. There was only a small variation in effective strain, and the hardness was generally homogeneous in the radial middle zones of the disks; however, a large variation occurred at the edge of the disk. Two lines were compared to determine the distribution of effective strain and are presented in **Figure 26**.

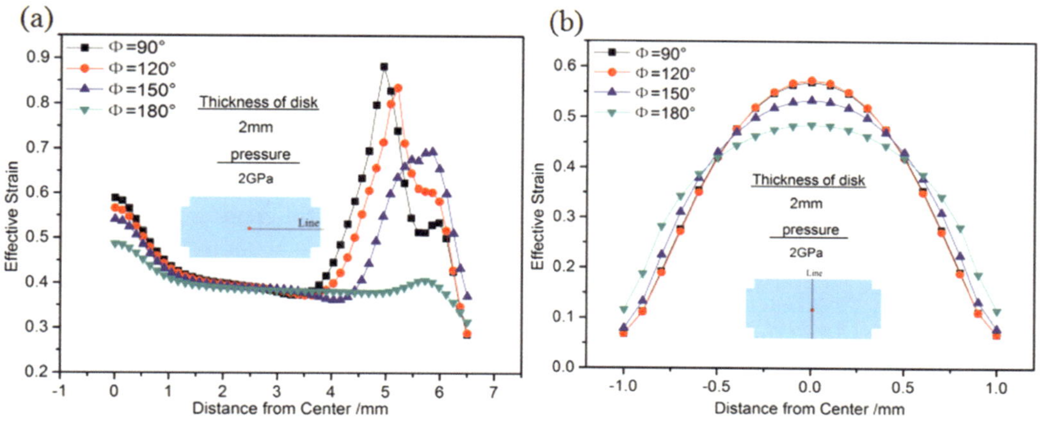

Figure 26. Path plots of the effective strain on lines listed in the disk at four wall angles of the cavity.

Figure 26 indicates that the plastic deformation was inhomogeneous along the thickness direction of the disk. Although the area of the hardness hill increased as the wall angle increased, the maximum effective strain in the area of the hardness hill decreased as the angle increased, as shown in **Figure 26b**. There was significant plastic deformation when the wall angle was less than 180° (indicated in **Figure 26a**), while the plastic deformation was relatively uniform along the radial direction under the wall angle condition of 180°. In addition, the anvil geometry also affected the flash of the disk. Interestingly, there was little variation in the flash length of the compressed disk at different wall angles, although a large distinction of strain in the flash regions happened. In any case, the wall angle of the cavity plays an important role in the heterogeneous deformation of the main body of the workpiece.

5. Effect of Hollowcone high-pressure torsion on the thermal and mechanics properties of Zr-based bulk metallic glass

Bulk metallic glasses (BMGs) usually have many outstanding properties, for example, good corrosion resistance [57], high elastic limits [58], high strength [59], and so on. However, the limited plastic deformability at room temperature is a problem awaiting solution, which seriously hindered the application of BMGs. However, BMGs under the condition of HPT exhibit a certain plasticity deformability at room temperature. The results from literatures indicated that the hardness of HPT BMGs disks decreased due to the rejuvenated structure, which is completely different phenomenon in the crystalline alloys. There also have some research reported that effects on the thermal properties of BMGs happened. For example, Edalati et al. reported that the glass transition temperature of $Zr_{50}Cu_{30}Ni_{10}Al_{10}$ BMG increased few after a HPT process [60]. However, Meng et al. reported that HPT did not change the glass transition temperature [61].

In particular, Kim et al. proposed that a hollow cone high-pressure torsion (HC-HPT) method used for the fabrication of hollow cone-shaped specimens with a closed cone head through the application of high pressure and torsion in order to obtain UFG/NC microstructures [62]. Therefore, it is an interesting and necessary attempt to research the effects of hollow cone high-pressure torsion on the performance of BMGs.

In this section, $Zr_{64.13}Cu_{15.75}Ni_{10.12}Al_{10}$ BMGs [63] were selected as a specimen material to research the effect of hollow-cone high-pressure torsion on the performance of BMGs. Because Zr-based BMGs have satisfactory glass-forming abilities, it can be undemanding cast into a hollow cone-shaped specimen. Furthermore, the catastrophic failure at room temperature will not occur by the HC-HPT process due to superior plastic deformability at room temperature.

Hóbora et al. reported that $Cu_{60}Zr_{20}Ti_{20}$ BMGs occurred crystallization after HPT process [64]. On the contrary, some research data indicate that BMGs did not occurred crystallization after HPT process. Therefore, first make sure whether the sample occurred crystallization after HC-HPT process. **Figure 27** shows the X-ray diffraction (XRD) patterns for the transverse cross sections of as-cast and HC-HPT 1 revolution BMG alloys.

The XRD patterns have similar profiles. The broad diffraction peaks indicate full vitrification of samples, and crystalline peaks were not detected. This indicates that no crystallization did occur in the $Zr_{64.13}Cu_{15.75}Ni_{10.12}Al_{10}$ BMG through HC-HPT 1 revolution. The plastic deformation during the HC-HPT process was reduced, conical wall thickness slight due to the sample has been reversed only one revolution.

HPT process changes the microstructure of the BMGs, even though have not induced crystallization, and inevitably have a certain effect on its thermodynamic properties. **Figure 28** presents the DSC profiles for the as-cast and HC-HPT 1 revolution samples.

The curves have similar profiles. They have well-defined glass transition regions. However, the characteristic temperatures of glass transition and crystallization are different. The Figure shows that the glass transition begins at 657.63 K and ends at 736.23 K for the as-cast sample, and it begins at 663.19 K and ends at 759.19 K for the HC-HPT 1 revolution sample. It is found that, after HC-HPT, supercooled liquid region increases.

The effects of HC-HPT process on the glass transition and crystallization behaviors for the Zr-based BMG are remarkable. The results happened by the increase of free volume after the large shear deformation of HC-HPT. The increase of free volume usually leads to a decrease in hardness and elastic modulus.

Figure 29 presents the curves of thermal expansion for the as-cast and HC-HPT 1 revolution processed at a heating rate of 0.0833 K/s. These curves exhibit similar thermal expansion behaviors. From the room temperature to around 640 K, the length of the both samples increased linearly to a temperature. Then, the lengths have decreased which suggest related to the crystallization of the alloys. Obviously, the average thermal expansion coefficient has

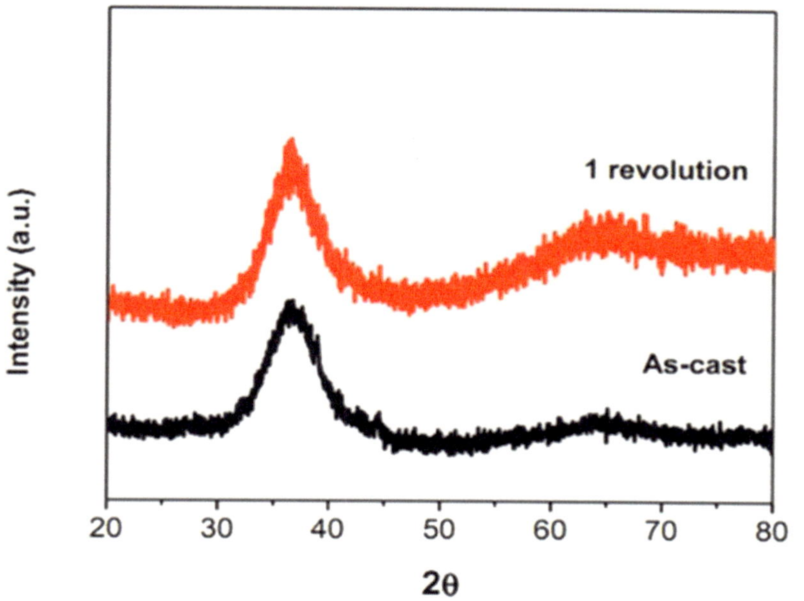

Figure 27. The XRD pattern of samples before and after deformation.

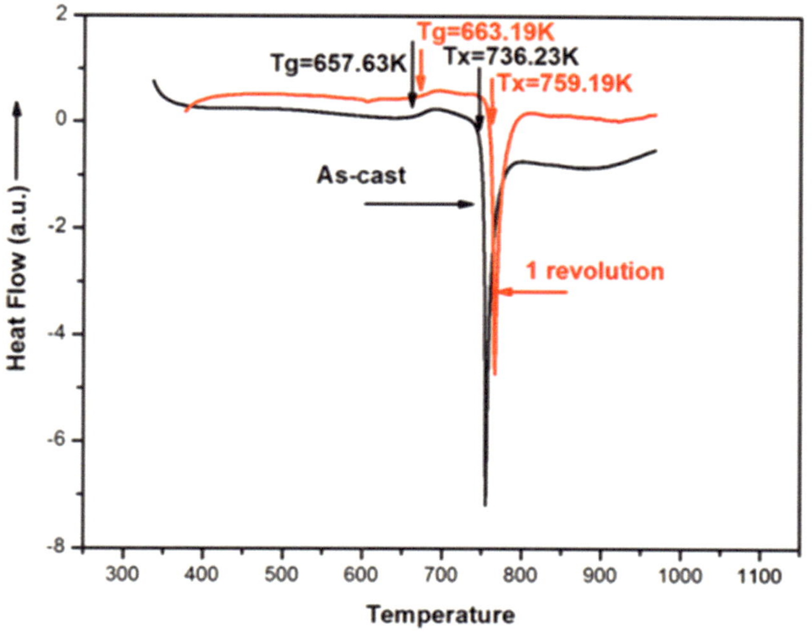

Figure 28. The DSC pattern of samples before and after deformation.

decreased after the HC-HPT process. The free volume can be created by the inelastic deformation of local atomic clusters under shear stress [65]. After the HC-HPT process, the free volume in the sample increased. Therefore, the sample was heated, and the created free volume by HC-HPT processed disappears due to a decrease in the thermal expansion coefficient.

Tan et al. [66] reported that the plasticity deformation corresponds to the internal states with more free volume as revealed by lower hardness and elastic modulus. **Figure 30** presents the hardness curve from cone vertical top to bottom.

The two patterns have similar profile, and the hardness of sample cone is tending to descend from top to bottom of the hollow cone. Compared with the sample before deformation, the hardness has decreased after HC-HPT. The results were coinciding with the increase of free volume. The same results were obtained for the elastic modulus test as shown in **Table 2**.

Table 2 presents the results using nano indentation method of elastic modulus of specimen at different position. And compared with HC-HPT 1 revolution sample, HC-HPT deformation makes the elastic modulus of amorphous alloy reduces. Different parts have different elastic modulus with a sample due to the special sample shape give rise to different cooling rates. A larger amount of free volume around atoms may enlarge their internal atomic spacing, decrease the atomic bonding strength, and thus lower the elastic modulus of the amorphous alloy [67].

In this study, hollow cone Zr-based amorphous has been chosen to be the investigated. Severe plastic deformation using HC-HPT with one revolution has effects on the thermal and mechanical properties of $Zr_{64.13}Cu_{15.75}Ni_{10.12}Al_{10}$ BMGs. It was found that after the HC-HPT process, the glass transition temperature of Zr-based BMG increased, but the Vickers hardness, elastic modulus, and coefficient of thermal expansion decreased.

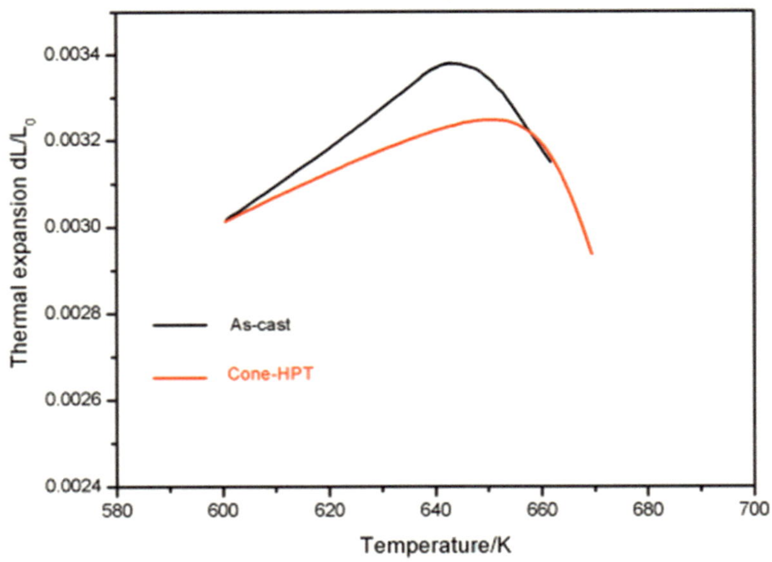

Figure 29. Dilatometer traces of the as-cast and cone-HPT processed.

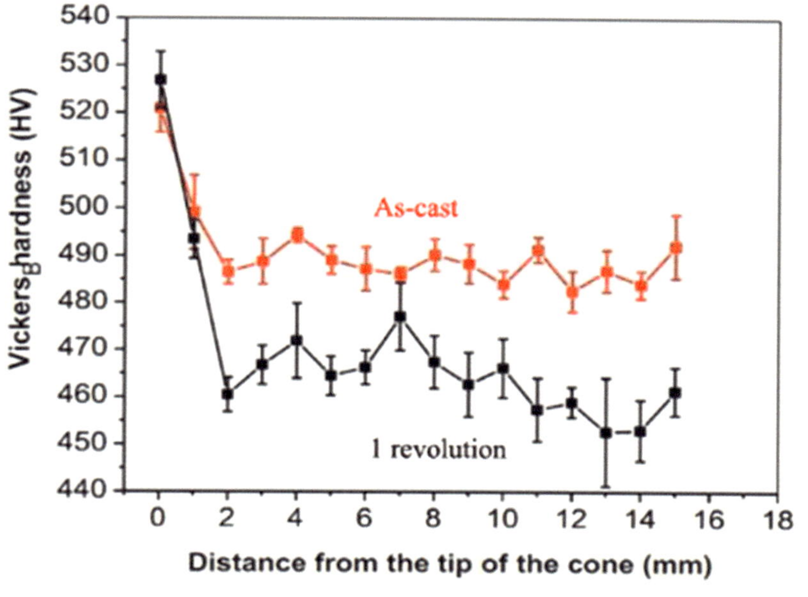

Figure 30. Vickers hardness of samples at different locations before and after HC-HPT.

Samples	E of top part (GPa)	E of middle part (GPa)	E of bottom part (GPa)
As-cast	95.16	92.92	82.43
One revolution	93.76	89.67	76.23

Table 2. Elastie modulus of specimens at different locations before and after HC-HPT.

Author details

Yuepeng Song[1]*, Wenke Wang[1,2], Miaomiao Chen[1], Jing Guo[1], Lingfeng Xu[1], Dongsheng Gao[1] and Hyoung Seop Kim[3]

*Address all correspondence to: ustbsong@sina.com

1 Shandong Provincial Key Laboratory of Horticultural Machineries and Equipments, Mechanical and Electronic Engineering College, Shandong Agricultural University, Tai'an, PR China

2 School of Materials Science and Engineering, Harbin Institute of Technology, Weihai, PR China

3 Department of Materials Science and Engineering, Pohang University of Science and Technology, Pohang, Korea

References

[1] Xie ZL, Xie JJ, Hong YS, Wu XL. Influence of processing temperature on microstructure and microhardness of copper subjected to high-pressure torsion. Science China Technological Sciences. 2010;**53**:1534-1539

[2] Kim HS, Estrin Y, Bush MB. Plastic deformation behaviour of fine-grained materials. Acta Materialia. 2000;**48**(2):493-504

[3] Kim HS, Estrin Y. Ductility of ultrafine grained copper. Applied Physics Letters. 2001; **79**:4115-4117

[4] Xu Ch, Horita ZJ, Langdon TG. The evolution of homogeneity in an aluminum alloy processed using high-pressure torsion. Acta Materialia. 2008;**56**:5168-5176

[5] Zhilyaev AP, Langdon TG. Using high-pressure torsion for metal processing: Fundamentals and applications. Progress in Materials Science. 2008;**53**:893-895

[6] Moon S, Kim JH, Shim JY, Ahn YB, Song KH. Analysis of aberrantly spliced HRPT2 transcripts and the resulting proteins in HPT-JT syndrome. Molecular Genetics and Metabolism. 2010;**100**:365-371

[7] Zhilyaev AP, Gimazov AA, Soshnikova EP, Révész Á, Langdon TG. Microstructural characteristics of nickel processed to ultrahigh strains by high-pressure torsion. Materials Science and Engineering A. 2008;**489**:207-212

[8] Zhilyaev AP, Oh-ishi K, Langdon TG, McNelley TR. Microstructural evolution in commercial purity aluminum during high-pressure torsion. Materials Science and Engineering A. 2005;**410-411**:277-280

[9] Lugo N, Llorca N, Cabrera JM, Horita Z. Microstructures and mechanical properties of pure copper deformed severely by equal-channel angular pressing and high pressure torsion. Materials Science and Engineering A. 2008;**477**:366-371

[10] Kurmanaeva L, Ivanisenko Y, Markmann J, Kübel C, Chuvilin A, Doyle S, Valiev RZ, Fecht HJ. Grain refinement and mechanical properties in ultrafine grained Pd and Pd–Ag alloys produced by HPT. Materials Science and Engineering A. 2010;**527**:1776-1783

[11] Kawasaki M, Ahn B, Langdon TG. Effect of strain reversals on the processing of high-purity aluminum by high-pressure torsion. Journal of Materials Science. 2010 (online)

[12] Zhilyaev AP, McNelley TR, Langdon TG. Evolution of microstructure and microtexture in fcc metals during high-pressure torsion. Journal of Materials Science. 2007;**42**:1517-1528

[13] Valiev RZ, Islamgaliev RK, Alexandrov IV. Bulk nanostructured materials from severe plastic deformation. Progress in Materials Science. 2000;**45**:103-189

[14] Yoo SJ, Kim WJ. Microstructure and strengthening mechanisms of carbon nanotube reinforced magnesium matrix composites fabricated by accumulative roll bonding. Korean Journal of Metals and Materials. 2014;**52**:561-172

[15] Valiev RZ, Langdon TG. Principles of equal-channel angular pressing as a processing tool for grain refinement. Progress in Materials Science. 2006;**51**:881-981

[16] Lee SH, Saito Y, Tsuji N, Utsunomiya H, Sakai T. Role of shear strain in ultragrain refinement by accumulative roll-bonding (ARB) process. Scripta Materialia. 2002;**46**:281-285

[17] Takayama Y, Sasaki J, Kato H, Watanabe H. Improvement of fatigue properties by means of continuous cyclic bending and annealing in an Al-Mg-Mn alloy sheet. Materials Transactions. 2004;**45**:1833-1838

[18] Dong HS, Kim I, Kim J, Zhu YT. Shear strain accommodation during severe plastic deformation of titanium using equal channel angular pressing. Materials Science and Engineering A. 2002;**334**:239-245

[19] Pourrahimi S, Nayeb Hashemi H, Foner S. High-strength high-conductivity Cu-Nb microcomposite wire by powder metallurgy. Journal of Materials Science Letters. 1990; **9**:1484-1487

[20] Prangnell PB, Harris C, Roberts SM. Finite element modeling of equal channel angular extrusion. Scripta Materialia. 1997;**37**:983-989

[21] Chang SY, Lee JG, Park KT, Dong HS. Microstructures and mechanical properties of equal channel angular pressed 5083 Al alloy. Materials Transactions. 2001;**42**:1074-1080

[22] Nakamura K, Neishi K, Kaneko K, Nakagaki M, Horita Z. Development of severe torsion straining process for rapid continuous grain refinement. Materials Transactions. 2004;**45**:3338-3342

[23] Orlov D, Beygelzimer Y, Synkov S, Varyukhin V, Tsuji N, Horita Z. Microstructure evolution in pure Al processed with twist extrusion. Materials Transactions. 2009;**50**:96-100

[24] Bridgman PW. Effects of high shearing stress combined with high hydrostatic pressure. APS Journals Archive. 1935;**48**:825-847

[25] Valiev RZ, Krasilnikov NA, Tsenev NK. Plastic deformation of alloys with submicron-grained structure. Materials Science and Engineering:A. 1991;**137**:35-40

[26] Figueiredo RB, Aguilar MT, Cetlin PR, Langdon TG. Deformation heterogeneity on the cross-sectional planes of a magnesium alloy processed by high-pressure torsion. Metallurgical and Materials Transactions A. 2011;**42**:3013-3021

[27] Hohenwarter A, Bachmaier A, Gludovatz B, Scheriau S, Pippan R. Technical parameters affecting grain refinement by high pressure torsion. International Journal of Materials Research. 2009;**100**:1653-1661

[28] Kim HS, Joo SH, Jeong HJ. Plastic deformation and computer simulations of equal channel angular pressing. Korean Journal of Metals and Materials. 2014;**52**:87-99

[29] Kim HS, Ryu WS, Janecek M, Baik SC, Estrin Y. Effect of equal channel angular pressing on microstructure and mechanical properties of IF steel. Advanced Engineering Materials. 2005;**7**:43-45

[30] Bridgman PW. On torsion combined with compression. Journal of Applied Physics. 1943;**14**:273-281

[31] Sakai G, Nakamura K, Horita Z, Langdon TG. Developing high-pressure torsion for use with bulk samples. Materials Science and Engineering A. 2005;**406**:268-273

[32] Song YP, Wang WK, Lee DJ, Jeong HJ, Lee S, Kim HS. Thickness inhomogeneity in hardness and microstructure of copper after the compressive stage in high-pressure torsion. Metals and Materials International. 2015;**21**:7-13

[33] Xu Ch, Horita ZJ, Langdon TG. Microstructural evolution in pure aluminum in the early stages of processing by high-pressure torsion. Materials Transaction. 2010;**51**:2-7

[34] Xu Ch, Horita ZJ, Langdon TG. The evolution of homogeneity in processing by high-pressure torsion. Acta Materialia. 2007;**55**:203-212

[35] Hadzima B, Janecek M, Estrin Y, Hyoung SK. Microstructure and corrosion properties of ultrafine-grained interstitial free steel. Materials Science and Engineering A. 2007;**462**:243-247

[36] Song YP, Wang WK, Gao DS, Kim HS, Yoon EY, Lee DJ, Lee CS, Guo J. Inhomogeneous hardness distribution of high pressure torsion processed IF steel disks. Materials Sciences and Applications. 2012;**3**:234-239

[37] Song YP, Wang WK, Gao DS, Yoon EY, Lee DJ, Kim HS. Finite element analysis of the effect of friction in high pressure torsion. Metals and Materials International. 2014;**20**:445-450

[38] Wang WK, Song YP, Yoon EY, Lee DJ, Lee JH, Kim HS. Analysis of stress states in the compression stage of high pressure torsion using the slab analysis method and the finite element method. Metals and Materials International. 2013;**5**:1021-1027

[39] Song YP, Wang WK, Gao DS, Yoon EY, Lee DJ, Lee CS, Kim HS. Hardness and microstructure of interstitial free steels in the early stage of the high-pressure torsion process. Journal of Materials Science. 2013;**48**:4698-4704

[40] Song YP, Yoon EY, Lee DJ, Lee JH, Kim HS. Mechanical properties of copper after compression stage of high-pressure torsion. Materials Science and Engineering A. 2011;**528**:4840-4844

[41] http://www.matweb.com

[42] Edalati K, Fujioka T, Horita Z. Microstructure and mechanical properties of pure Cu processed by high-pressure torsion. Materials Science and Engineering A. 2008;**497**:168-173

[43] Wang WK, Song YP, Kim HS. Experimental and finite element analyses of plastic deformation in the compressive stage of copper in high-pressure torsion. Rare Metal Materials and Engineering. 2011;**40**:301-304

[44] Vorhauer A, Pippan R. On the homogeneity of deformation by high pressure torsion. Scripta Materialia. 2004;**51**:921-925

[45] Figueiredo RB, Langdon TG. Fabricating ultrafine-grained materials through the application of severe plastic deformation: A review of developments in Brazil. Journal of Materials Research and Technology. 2012;**1**:55-62

[46] Bachmaier A, Kerber M, Setman D, Pippan R. The formation of supersaturated solid solutions in Fe-Cu alloys deformed by high-pressure torsion. Acta Materialia. 2012;**60**:860-871

[47] Xum C, Horita Z, Langdon TG. Evaluating the influence of pressure and torsional strain on processing by high-pressure torsion. Journal of Materials Science. 2008;**43**:7286-7292

[48] Kawasaki M, Langdon TG. The significance of strain reversals during processing by high-pressure torsion. Materials Science and Engineering: A. 2008;**498**:341-348

[49] Yoon SC, Nagasekhar AV, Kim HS. Finite element analysis of the bending behavior of a workpiece in equal channel angular pressing. Metals and Materials International. 2009;**15**:215-219

[50] Ivanisenko Y, Lojkowski W, Valiev RZ, Fecht HJ. The mechanism of formation of nanostructure and dissolution of cementite in a pearlitic steel during high pressure torsion. Acta Materialia. 2003;**51**:5555-5570

[51] Horita Z, Langdon TG. Achieving exceptional superplasticity in a bulk aluminum alloy processed by high-pressure torsion. Scripta Materialia. 2008;**58**:1029-1032

[52] Shen G, Vedhanayagam A, Kropp E. A method for evaluating friction using a backward extrusion-type forging. Journal of Materials Processing Technology. 1992;**33**:109-123

[53] Narayanan RG, Gopal M, Rajadurai A. Influence of friction in simple upsetting and prediction of hardness distribution in a cold forged product. Journal of Testing and Evaluation. 2008;**36**:371-383

[54] Xu SQ, Chen ZY, Zhang SY. Research on the relation between bulging and friction in cylinder upsetting. Forging and Stamping Technology(in Chinese). 2004;**29**:46-48

[55] Lu P, Zhao GQ, Guan YJ, Cheng JL. Thickness of decarburized Layer can be calculated by Fick's Law of diffusion. China Metal Forging Equipment and Manufacturing Technology (in Chinese). 2008;**43**:75-79

[56] Lee DJ, Yoon EY, Park LJ, Kim HS. The dead metal zone in high-pressure torsion. Scripta Materialia. 2012;**67**:384-387

[57] Li YH, Zhang W, Qin FX, Makino A. Mechanical properties and corrosion resistance of a new $Zr_{56}Ni_{20}Al_{15}Nb_4Cu_5$ bulk metallic glass with a diameter up to 25mm. Journal of Alloys and Compounds. 2014;**615**:71-74

[58] Zong HT, Bian LY, Cheng JY, Cao GH, Kang CY, Li M. Glass forming ability, thermal stability and elastic properties of Zr-Ti-Cu-Be-(Fe) bulk metallic glasses. Results in Physics. 2016;**6**:1157-1160

[59] Zhou XL, Zhou HF, Li XY, Chen CQ. Size effects on tensile and compressive strengths in metallic glass nanowires. Journal of the Mechanics and Physics of Solids. 2015;**84**:130-144

[60] Edalati K, Yokoyama Y, Horita Z. High-pressure torsion of machining chipsand bulk discs of amorphous $Zr_{50}Cu_{30}Al_{10}Ni_{10}$. Materials Transactions. 2010;**51**:23-26

[61] Meng FQ, Tsuchiya SK, Yokoyama Y. Reversible transition of deformation mode by structural rejuvenation and relaxation in bulk metallic glass. Applied Physics Letter. 2012;**101**:121914

[62] Um HY, Yoon EY, Lee DJ, Lee CS, Park LJ, Lee S, Kim HS. Hollow cone high-pressure torsion: Microstructure and tensile strength by unique severe plasticdeformation. Scripta Materialia. 2014;**71**:41-44

[63] Liu YH, Wang G, Wang RJ, Zhao DQ, Pan MX, Wang WH. Super plastic bulk metallic glassesat room temperature. Science. 2007;**315**:1385-1388

[64] Hóbora S, Kovács Z, Révész Á. Estimation of heat production during high pressure torsion of Cu-based metallic glass. Journal of Alloys and Compounds. 2010;**495**:352-355

[65] Steif PS, Spaepen F, Hutchinson JW. Strain localization in amorphous metals. Acta Metallurgica. 1982;**30**:447-455

[66] Tan J, Zhang Y, Sun BA, Stoica M, Li CJ, Song KK, Kühn U, Pan FS, Eckert J. Correlation between internal states and plasticity in bulk metallic glass. Applied Physics Letters. 2011;**98**:15

[67] Li N, Liu L, Chen Q, Pan J, Chan KC. The effect of free volume on the deformation behaviour of a Zr-based metallic glass under nanoindentation. Journal of Physics D: Applied Physics. 2007;**40**:19

Chapter 2

Mechanical Properties and Microstructure Development in Ultrafine-grained Materials Processed by Equal-channel Angular Pressing

Peter Minárik, Tomáš Krajňák, Ondřej Srba,
Jakub Čížek, Jenő Gubicza, Milan Dopita,
Radomír Kužel and Miloš Janeček

Additional information is available at the end of the chapter

http://dx.doi.org/10.5772/intechopen.68965

Abstract

In this chapter, the detailed characterization of processes of grain fragmentation and refinement resulting from gradual imposition of strain by individual equal-channel angular pressing (ECAP) passes is reported. A great emphasis is placed on the processing of materials with different crystal structure, particularly the face-centred cubic (FCC), the body-centred cubic (BCC) and the hexagonal close-packed (HCP). Advanced techniques of electron microscopy, electron and X-ray diffraction and positron annihilation spectroscopy have been employed to characterize microstructure, texture and defect evolution in the material as a function of strain imposed by ECAP. Microstructure development was correlated with mechanical properties obtained by both mechanical tests and microhardness measurements. Processes controlling the microstructure refinement and texture development were identified and discussed in detail.

Keywords: ECAP, microstructure, crystal structure, mechanical properties, dislocations

1. Introduction

Ultrafine-grained (UFG) materials processed by severe plastic deformation exhibit enhanced mechanical, electrical, corrosion, magnetic and other physical properties [1]. Several techniques of grain refinement imposing severe plastic deformation (SPD) into the material have been developed in the last several decades. Among them, equal-channel angular pressing (ECAP) has been the most widely used. A large variety of materials—pure metals, alloys, composites, etc.—with different crystal structure have been successfully processed by ECAP. Usually, a bar- or rod-shaped sample is pressed through a channel in a die, which is bent in a sharp angle where the pure shear occurs (see **Figure 1**).

The cross-section of the sample is unchanged after the processing, and therefore it can be repeatedly processed to obtain high degree of strain. Additionally, there is a possibility to activate different slip systems after each pass by rotating the sample along its processing direction. There are four basic processing routes which describe the way of the billet rotation between individual processing passes. Route 'A' corresponds to no billet rotation, while routes 'B_C' and 'B_A' refer to the rotation by 90° in the same direction and alternate direction, respectively, and route 'C' refers to the rotation by 180° after each pass [3]. The attractiveness of ECAP increases its scalability, which enables the processing of both very small and large samples. Homogeneity of the microstructure is not affected by the size of the sample, and therefore ECAP is a unique technique, which enables production of larger bulk UFG material,

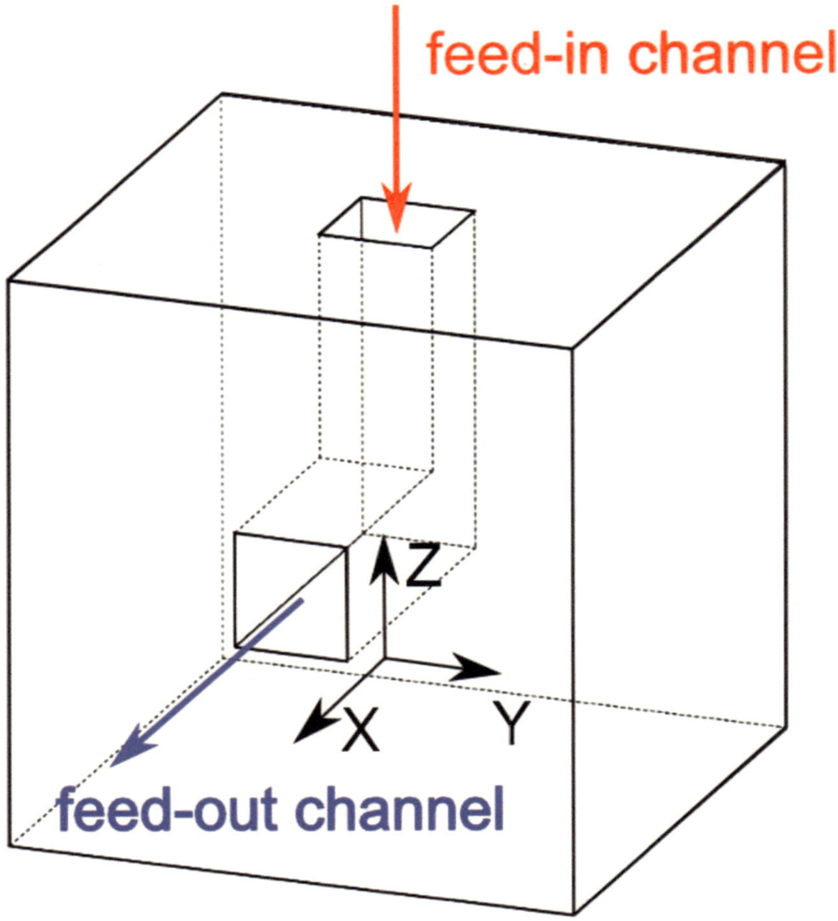

Figure 1. The scheme of the ECAP die including the coordinate system [2].

when compared to other SPD techniques. Additionally, continual ECAP, denoted as ECAP-conform, has been developed to process very long sheets and wires [4].

2. Experimental techniques

In this section, the experimental techniques which were employed to characterize microstructure and crystal lattice defect evolution and mechanical properties of the ultrafine-grained materials after ECAP are introduced.

X-ray diffraction line profile analysis (XLPA) and positron annihilation spectroscopy (PAS) are indirect, non-destructive but powerful methods for the characterization of defect structure of ultrafine-grained materials and nanomaterials. The PAS technique is based on the measurement of lifetimes of positrons in irradiated material where the positrons are trapped preferentially on defects of the crystal lattice with lower electron density, which increases effectively their lifetime, with respect to that of free positrons. Detailed information about the principle of PAS is presented elsewhere [5]. The concentration of defects was calculated from PAS parameters using a diffusion trapping model (DTM), assuming non-uniform spatial distribution of defects (dislocations, vacancies, vacancy clusters, etc.) [6, 7]. In this model, the sample is considered as a combination of almost dislocation-free cells having a diameter (d) and dislocation walls containing extremely high dislocation density. The mean dislocation density (ρ) is calculated from the volume fraction of dislocation walls (η), and the dislocation density in the dislocation walls ($\rho_{D,wall}$).

In XLPA technique, the defect structure parameters are derived from the analysis of the width and shape profiles of diffraction peaks. The line broadening due to the strain induced by dislocations and the crystallite size are different in nature, allowing separation of their contributions easily. The results presented in this chapter were obtained from the evolution of high-resolution X-ray diffraction patterns by the convolutional multiple whole profile (CMWP) fitting method [8]. This procedure is based on fitting of the diffraction pattern by the convolution of the background spline, the instrumental pattern and theoretical line profiles reflecting the real structure (crystallite size and dislocations). The theoretical size profile function is modelled by the microstructure consisting of spherical crystallites and the log-normal distribution. The dislocation distribution in individual slip systems is determined by comparison of experimental contrast factors of dislocations and the theoretical values employing the procedure described in detail elsewhere [9].

PAS and XLPA spectroscopic techniques were complemented by direct observations of UFG structure by means of electron microscopy and electron diffraction. Diffraction contrast in the transmission electron microscope (TEM) was employed for the observation of dislocation and subgrain structure [10]. Crystal orientation maps were obtained by electron backscatter diffraction (EBSD) [11]. Measured EBSD data were evaluated by software TSL OIM Analysis 7, which enables to determine the crystallographic microtexture, the character and the fraction of grain boundaries and to evaluate the mean grain size. The macrotexture measurements were performed by X-ray diffraction.

Mechanical behaviour of the samples was tested both under the uniaxial tension with a constant strain rate and by Vickers microhardness measurement, which simulates multi-axial loading.

3. Materials with FCC structure

3.1. Material processing and experimental procedure

Copper of technical purity having the following content of alloying elements (Fe, 0.0037, $P < 0.001$; Sb < 0.0003; Bi < 0.0001; As < 0.001, in wt%) as a typical representative of the material with FCC structure was used in this study. The material received initially as cast condition underwent the homogenization annealing at 450°C for 4 hours. A series of billets of the diameter of 10 × 10 mm and the length of 80 mm were processed by ECAP at room temperature for different numbers of passes N ($N = 1$, 2, 4 and 8) following route B_C.

Microstructure evolution with strain imposed on the material by different numbers of ECAP passes was investigated by electron microscopy, including EBSD and X-ray diffraction. The evolution of lattice defects (dislocations and vacancies) was investigated by XRD and positron annihilation spectroscopy. Microstructure development was correlated with mechanical properties obtained by tensile tests at room temperature.

3.2. Microstructure evolution

The microstructure of the material in the initial condition is shown in **Figure 2**. It consists of fully recrystallized grains of the average size of approximately 50–100 μm. Numerous twins resulting from the homogenization annealing are also seen at the light micrograph in **Figure 2**.

The microstructure evolution of Cu after different numbers of passes was observed by transmission electron microscopy (TEM). TEM specimens were prepared from the plane perpendicular to the pressing direction (plane X [12]). TEM micrographs displaying the characteristic structures after the individual number of passes are shown in **Figure 3**. It is seen that already after the first ECAP, the strong grain refinement occurs. The microstructure consists of elongated dislocation cells and/or subgrains. Two typical kinds of contrast may be distinguished at the micrograph, namely, dark lines corresponding to dense dislocation walls [13] and bright wider zones comprising individual subgrain boundaries. The comparison of diffraction pattern from both parts of grain boundaries indicates that the majority of grain boundaries have a low misorientation which is characteristic for so-called low-angle grain boundaries (LAGBs). The clear alignment of the structure along <111> direction is seen in the micrograph in **Figure 3(a)**. The overall character of the microstructure corresponds to the heavily deformed material. After the second pass of ECAP (**Figure 3b**), the microstructure remained almost unchanged. The cell/subgrain size was reduced only slightly (the average length of 800 nm and the average width of 200–300 nm) and structure remained aligned along <111> direction. A slight deviation of several subgrains from <111> indicates the activation of other slip systems during the second ECAP pass. After four passes of ECAP, significant

Figure 2. Microstructure of the non-deformed Cu.

changes in the microstructure occurred (see **Figure 3c**). The majority of grains are already equiaxed and the fraction of high-angle grain boundaries (HAGBs) characterized by typical thickness fringes contrast increased. The activation of new glide systems in planes which are not parallel with the original glide planes due to the rotation of the billet between individual passes must have occurred between the second and fourth pass. The typical microstructure of the specimen after eight passes is shown in **Figure 3(d)**. The microstructure is homogeneous consisting of equiaxed grains separated by sharp HAGBs. Significantly lower density of dislocations in grain interiors was observed in this condition. The average grain size ranged between 200 and 300 nm. In some zones, subgrains/grains with as large as about 500 nm were also observed. TEM observations confirm the efficient grain refinement of Cu polycrystals by ECAP (factor of 1000).

Inverse pole figures obtained from *EBSD measurements* are presented in **Figure 4**. The results are consistent with local TEM observations. Relatively large areas of boundaries with low misorientations indicated by slight colour code variations corresponding to bands subgrains observed by TEM (cf. **Figure 2**) dominate the microstructure [14]. With increasing the number of ECAP passes, much smaller subgrains and grains having larger misorientations evolve from these zones. A detailed analysis of grain boundary character distributions in individual specimens after ECAP is given elsewhere [15]. In specimens after four and eight passes, EBSD measurements revealed a high density of twin ($\sum 3$) and multiple twin boundaries $\sum 3^n$ ($\sum 9$, $\sum 27$), see **Figure 5**. Other authors also reported a similar result. This extensive formation of $\sum 3^n$ special boundaries is assumed to occur only after extensive strain hardening imposed by severe plastic deformation once a certain critical dislocation density

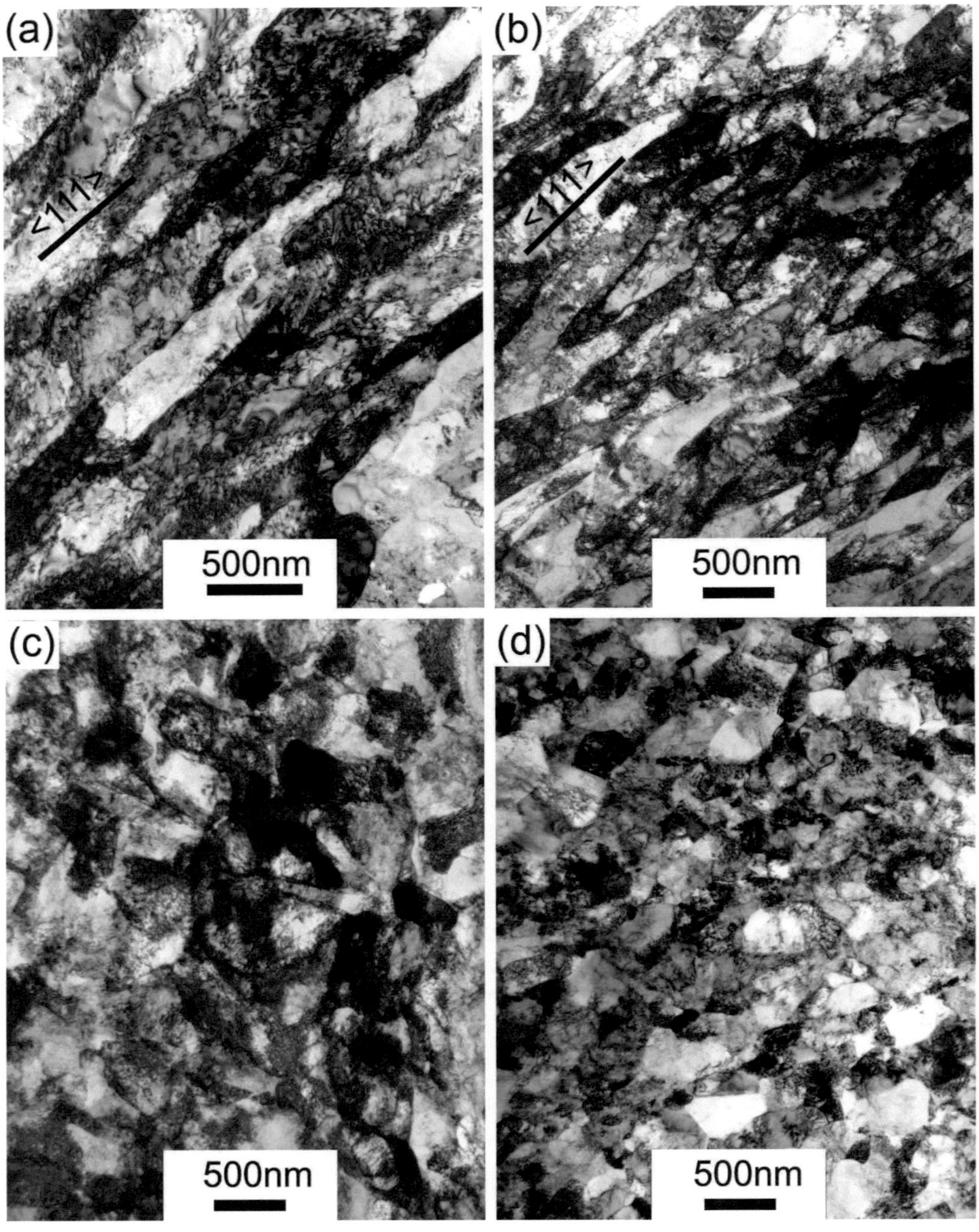

Figure 3. TEM micrographs of the typical microstructure of Cu specimens after different number of ECAP passes N (plane X) (a) $N = 1$, (b) $N = 2$, (c) $N = 4$ and (d) $N = 8$.

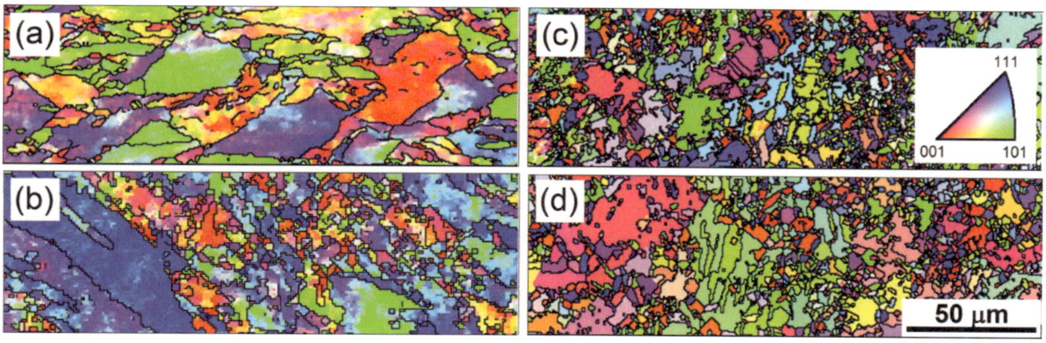

Figure 4. IPF map of ECAP specimens (a) 1P, (b) 2P, (c) 4P, (d) 8P (plane X).

is reached [16, 17]. EBSD also confirmed enhanced presence of LAGBs in the specimen after 1P (see **Figure 5**). With increasing strain due to ECAP, LAGBs ($\theta < 15°$) were continuously transformed into HAGBs ($\theta > 15°$). After eight passes, almost 90% of all grain boundaries had a high-angle character as displayed in **Figure 6**.

It is well known that the material subjected to severe plastic deformation contains high density of lattice defects, namely, dislocations and point defect. The evolution of density of lattice defects as a function of the number of ECAP passes was investigated by *positron*

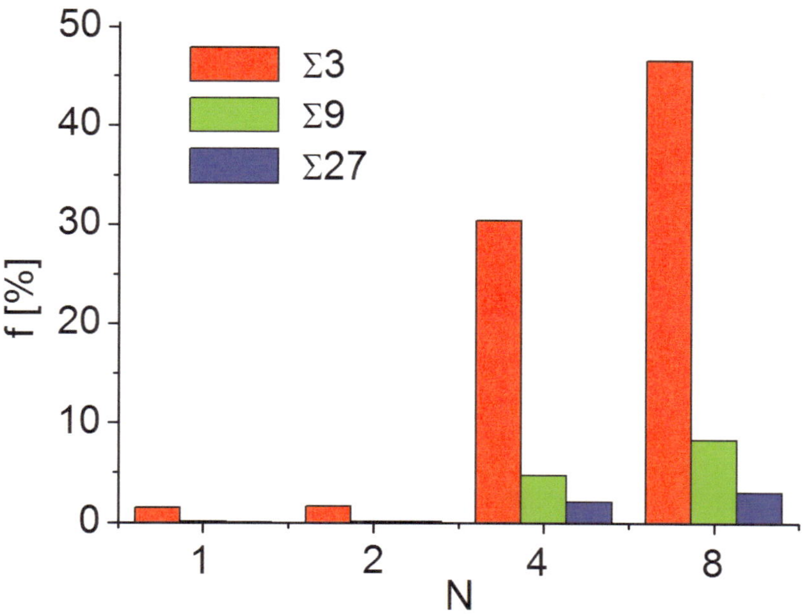

Figure 5. Special boundaries distribution.

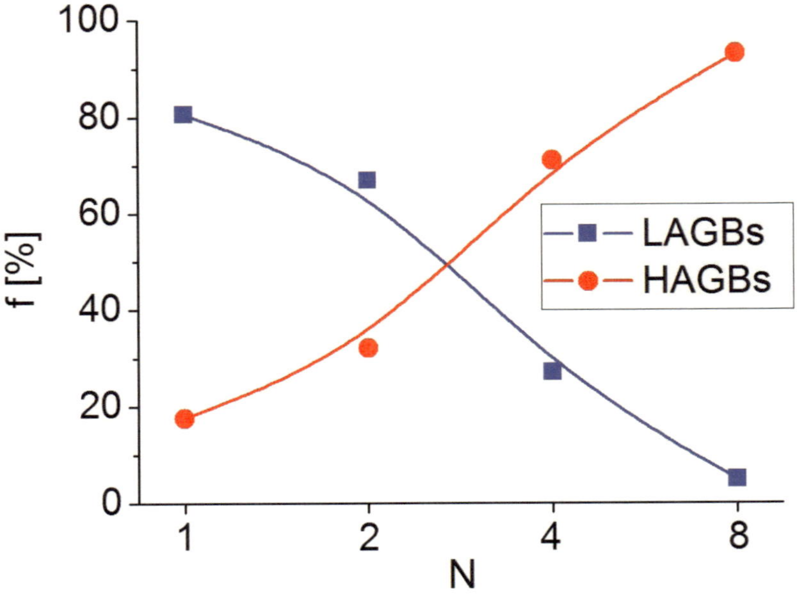

Figure 6. GB character evolution in ECAP Cu.

annihilation spectroscopy. Three different components were found in positron lifetime (PL) spectra as depicted in **Figure 7(a)**: (i) a component corresponding to free positrons which are delocalized in the lattice with the respective lifetime $\tau_1 \approx 114$ ps and the relative intensity I_1, (ii) a component representing the contribution of positrons trapped at dislocations [18] having a lifetime of $\tau_2 \approx 164$ ps and the intensity I_2, and (iii) a component attributed to positrons trapped at small vacancy clusters called microvoids formed by clustering of vacancies generated by ECAP [7] with the lifetime τ_3 depending on the number of vacancies in the cluster and the intensity I_3.

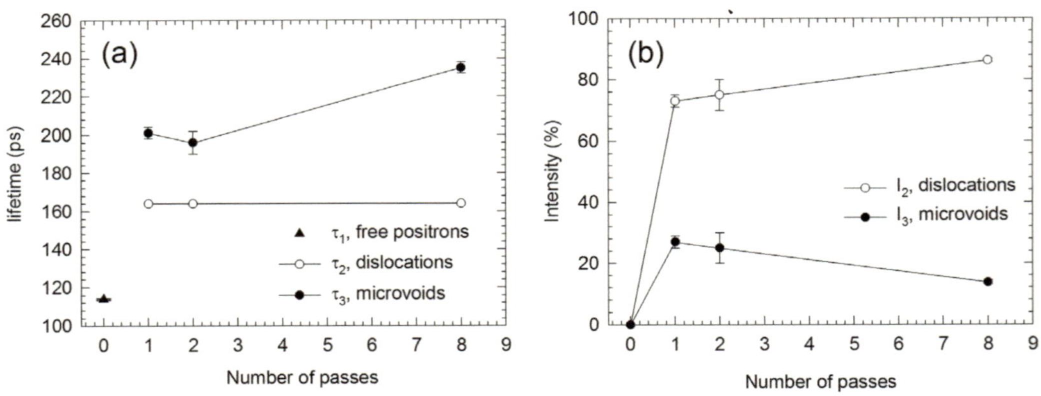

Figure 7. PAS results for Cu specimens subjected to various numbers of ECAP passes: (a) lifetimes of the components resolved in PL spectra, (b) intensities of the components arising from positrons trapped at defects.

http://dx.doi.org/10.5772/intechopen.68965

In annealed non-deformed Cu specimen ($N = 0$), only a single component PL spectrum having a lifetime of τ_1 = 114 ps, I_1 = 100%, was found (cf. **Figure 7a**). In this specimen, all positrons are annihilated from the free state. It should be noted that the lifetime τ_1 agrees well with the theoretically calculated lifetime of free positrons in Cu [19].

In all specimens deformed by ECAP, no free positron component was found $I_1 = 0\%$ (so-called saturated trapping). The evolution of lifetimes with N indicates that the component τ_2 is independent of N testifying that only the density of dislocations changes during ECAP pressing while the character of dislocation traps does not. On the other hand, the component τ_3 changes with increasing number of passes corresponding to the change of the microvoid size, i.e. the number of vacancies in the respective microvoids.

Due to the high density of lattice defects in ECAPed Cu specimens, almost all positrons are trapped at the open volume around these defects. In this case, the dislocation density ρ_D exceeds the value of 10^{14} m^{-2} and the microvoid concentration c_v the value of 10^{-4} at^{-1}. The intensity of positrons trapped at dislocations $I_2$was found to increase with increasing number of ECAP passes as shown in **Figure 7(b)**. On the other hand, the intensity of positrons trapped at microvoids I_3increases only during the first pass, while for $N > 1$ it decreases continuously up to $N = 8$.

Due to the saturated trapping of positrons in defects (I_1 = 0%), it is not possible to determine the absolute values of defect densities from the DTM model [5, 7]. Instead of that, only the ratio of the respective defect densities (ρ_D/c_v) may be determined from the ratio of respective intensities of positrons trapped at dislocations and at microvoids I_2/I_3 according to the formula (1) [7], where K_D and K_v are the trapping rates of positrons to dislocations and to microvoids, respectively. The variation of the ratio K_D/K_v with the number of ECAP passes is plotted in **Figure 8(a)**. The curve indicates that the dislocation density increases faster than the concentration of microvoids in all specimens deformed by ECAP.

$$\frac{I_2}{I_3} = \frac{K_D}{K_v} \sim \frac{\rho_D}{c_v} \tag{1}$$

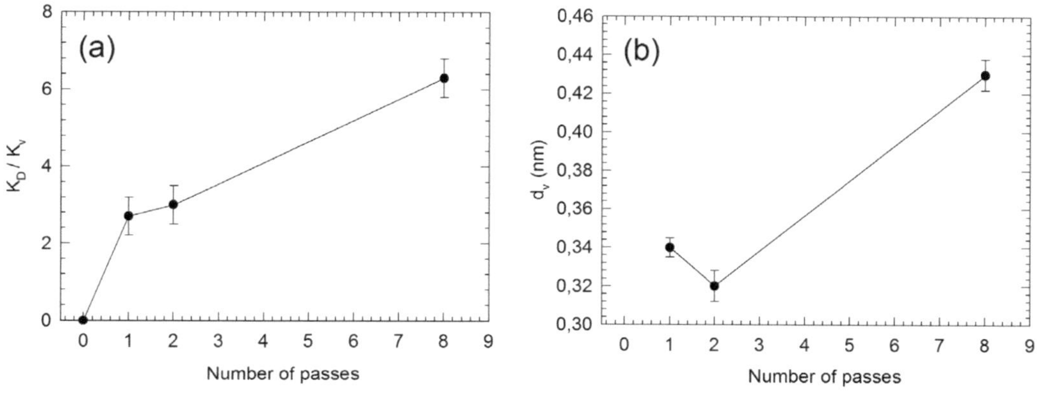

Figure 8. (a) The ratio K_D/K_v of positron trapping rate to dislocations and microvoids, (b) diameter of microvoids calculated from PAS results.

Theoretical calculations described in detail in Ref. [7] allow one to determine qualitatively the size of microvoids from the component $\tau_{3.}$ **Figure 7(b)**, which presents the dependence of the microvoid diameter d_v as a function of the number of passes, shows the results of these calculations. The microvoid diameter size change seems to be statistically insignificant in specimens deformed up to $N = 2$, while significantly larger microvoids (containing the higher number of monovacancies) are created in specimens deformed by eight ECAP passes.

The drawback of the PAS, which is unable to determine quantitatively the dislocation density in specimens containing high density of these defects ($\rho \geq 10^{14}$ m^{-2}, saturated trapping) [4, 6] is smeared by employing other technique, namely, *X-ray diffraction* (XRD), making use of the fact that peak broadening is caused both by the size of coherently scattering domains and by the presence of strain in the material. The former effect allows determining the size of these domains while the latter one allows to determine quantitatively the density of dislocation employing the techniques of the whole total patter fitting considering the size and dislocation broadening models proposed by Ribárik et al. [20] and Matěj et al. [21]. Extended FOX model proposed by Matěj et al. [21] was used to evaluate the results of XRD measurements in Cu specimens deformed by ECAP. This model uses 4 to 5 fitting parameters, namely, the dislocation density, the mean crystallite size, the dislocation correlation factor, the variance of the size distribution and/or the fraction of screw and edge dislocations, respectively. The details of fitting procedure are described in detail elsewhere [20]. The results of the fitting procedure are summarized in **Table 1**.

The density of dislocations ρ_D was found to increase with increasing number of ECAP passes which is in good agreement with qualitative predictions of PAS. Moreover, the fraction of edge dislocation decreases with increasing strain, which is consistent with results of other authors obtained on Cu [22]. This fact is attributed to the enhanced mobility of edge dislocations compared to screw ones [23]. The mean crystallite size determined by XRD is significantly lower than that obtained by TEM or EBSD. It is well known that crystallite size determined by XRD corresponds to the size of coherently scattering domains which reflects the size of subgrains rather than grains [24]. As a consequence, the values of <*L*> are significantly lower than the values of grain size determined by electron microscopy and electron diffraction <d> (see **Table 1**).

3.3. Mechanical properties

Mechanical properties were determined by testing tensile specimens in a universal screw-driven Instron 5882 machine at the initial strain rate of 4×10^{-4} s^{-1}at room temperature (RT).

Sample	ρ_D, 10^{15} m^{-2}	w	<L>, nm	<d>, nm
$N = 1$	2.1	0.95	78	900–700
$N = 2$	6.6	0.85	71	800–700
$N = 8$	7.9	0.35	76	150–250

Note: ρ_D—dislocation density; *w*—fractions of edge dislocations; <*L*>—the mean crystallite size;**<d>** , nm—the grain size determined by TEM/EBSD.

Table 1. Structure parameters for ECAP Cu samples for different numbers of passes $N = 1$, 2 and 8.

http://dx.doi.org/10.5772/intechopen.68965

The true stress-true strain curves for the non-deformed (0P) and ECAPed specimens (N = 1, 2, 4 and 8P) are shown in **Figure 9**. **Table 2** summarizes tensile test data in terms of the yield $\sigma_{0.2}$ and ultimate tensile strength σ_{max}(YS and UTS, respectively) and the total elongation (ε_{tot}).

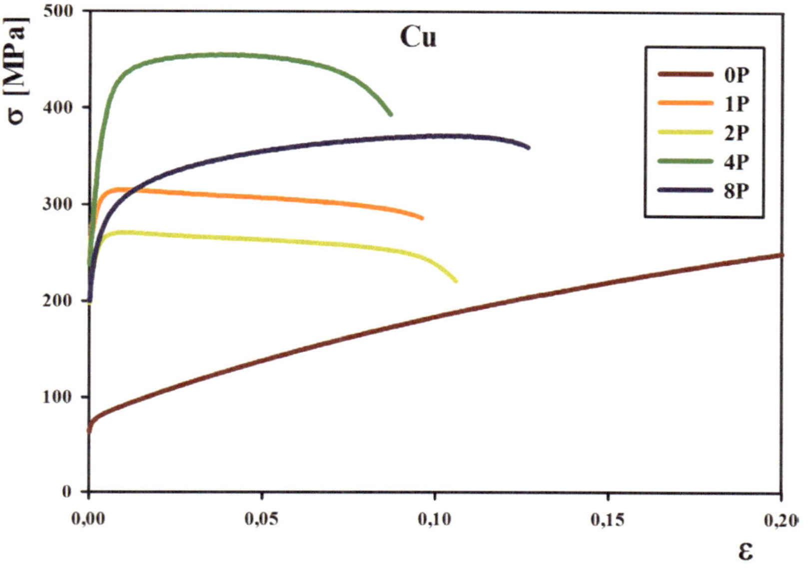

Figure 9. True stress vs. true strain curves of Cu specimens after different numbers of ECAP passes.

ECAPed specimens exhibit significantly better mechanical properties as compared to the coarse-grained material (N–0). Both YS and UTS increase up to four ECAP passes and slightly decline after eight passes. Mechanical properties correlate well with the microstructure evolution in individual specimens after ECAP. The increase of both YS and UTS up to four passes corresponds well to the combined effect of both the reduction in grain size and the increase of dislocation density. As no significant change of grain size was found in specimens N > 4, the mechanical properties in these specimens are influenced mainly by dislocation density variations. PAS (qualitatively) and XRD (quantitatively) measurements indicate that ρ_D increases up to N = 4 and is followed by a slight decline in the specimen N = 8 [25]. It is fully consistent with the evolution of mechanical properties with strain imposed by ECAP. However, the drop of YS and UTS in the specimen after

No. of passes/N	0	1	2	4	8
$\sigma_{0.2}$ (MPa)	78	293	250	330	258
σ_{max} (MPa)	215	314	270	455	371
ε_{tot} (%)	40	9.5	10.6	8.7	12.7

Table 2. Summary of experimental data obtained from mechanical testing.

eight passes may also be attributed to other microstructural effects, in particular the development of the grain boundary character distribution (cf. **Figures 4** and **5**) and the texture development which is currently under evaluation.

4. Materials with BCC structure

4.1. Material processing and experimental procedure

Single-phase ferritic interstitial-free (IF) steel having the BCC crystal structure and the carbon content less than 0.01 wt% was pressed through a rectangular ECAP die at room temperature (RT). ECAP billets were pressed for one, two, four and eight passes with the speed of 2 mm/min via route B_C. Microstructural characterization of samples was performed by conventional EBSD and TEM techniques. The lattice defects were studied by positron annihilation spectroscopy (PAS) employing the diffusion trapping model (DTM) [6, 7]. Mechanical properties at RT were characterized by Vickers microhardness measurement and the tensile tests with a constant strain rate of 10^{-3} s^{-1}. Detailed information about the composition, sample preparation and experimental methods can be found elsewhere [26–28].

4.2. Microstructure characterization by EBSD and TEM

Figure 10 shows the homogeneous microstructure of the initial state. The high-angle grain boundaries (HAGBs) (>15°) are outlined by black colour. The microstructure exhibits random crystallographic texture and is formed by equiaxed grains with a mean grain size of about 41 μm. The mean grain size was determined from the EBSD images as the area-weighted mean grain size and only grains separated by HAGBs were taken into account. Microstructural evolution during ECAP processing is displayed in **Figure 11**. Increasing the number of ECAP passes (*N*) leads to the gradual refinement of the microstructure. After the single ECAP pass, heavily deformed grains can be observed. Different colour inside the original grains indicates the formation of subgrains or dislocation cells having low-angle misorientations (LAGBs). After the second pass, the misorientation across the low-angle grain boundaries increases and some of them are transformed into the high-angle grain boundaries indicated by the black colour. With further straining, the gradual increase in the fraction of HAGBs can be observed. After 8 passes, almost fully refined microstructure with the average grain size of 0.7 μm and the fraction of HAGBs 65% was observed.

Detailed TEM observation confirmed the formation of bands of subgrains with sharp boundaries and dislocations cells with fuzzy boundaries in the sample after the single pass (see **Figure 12a**). After eight passes, new refined grains with the size around 500 nm are formed from subgrains in these deformation bands. As a consequence, the misorientation of refined grains remains low in same regions, i.e. they are separated predominantly by LAGBs. Additionally, a non-uniform spatial distribution of dislocations was observed: grain/subgrain interiors almost free of dislocations are separated by distorted layers with a very high density of dislocations. Mechanism of grain refinement described above was observed in many other materials with FCC and BCC crystal structure processed by ECAP [29–32].

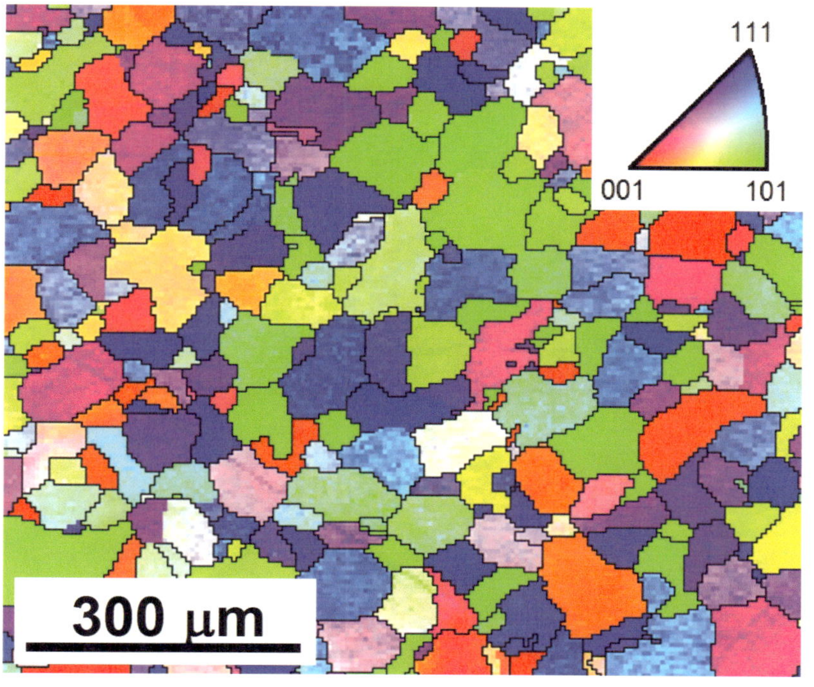

Figure 10. EBSD inverse pole figure map of the initial state (size-rolled and homogenized for 1 hour at 700°C).

Microtexture evolution during ECAP processing is described by a series of EBSD (100) and (110) pole figures measured on the plane X [12], lying perpendicular to the pressing direction (see **Figure 13**). With increasing number of passes, a gradual formation of three strong maxima in the EBSD (110) pole figure was observed. The maxima are tilted by 45° from each other. The analysis of interplanar angles in cubic crystals [33] indicates that these maxima are associated with {110} planes. After the eight passes, the stronger (110) texture in comparison with that after four passes is formed. This is consistent with the observations of De Messemaeker et al. [34]. The maxima are tilted roughly by 20° towards the y-axis. This is consistent with the cloud-model of Toth and co-workers [35, 36] and has been reported by other authors in BCC material [35–37].

4.3. Defect structure investigation by PAS

Lifetimes of the exponential components resolved in the lifetime (LT) spectra are plotted in **Figure 14(a)** as a function of the number of ECAP passes. The initial sample (0N) exhibits a single component spectrum with the lifetime of ≈108 ps, which can be attributed to the positrons annihilated in the free state. As a consequence, the initial sample exhibits a low density of defects (dislocation density below 5×10^{12} m^{-2}). The samples deformed by ECAP ($N > 0$) exhibit two-component LT spectra. Apart from the component τ_1 attributed to positrons not trapped at defects, the longer component with the lifetime $\tau_2 \approx 150$ ps coming from positrons trapped at dislocations [38–40] were detected in specimens $N > 0$. Park et al. [38] performed a

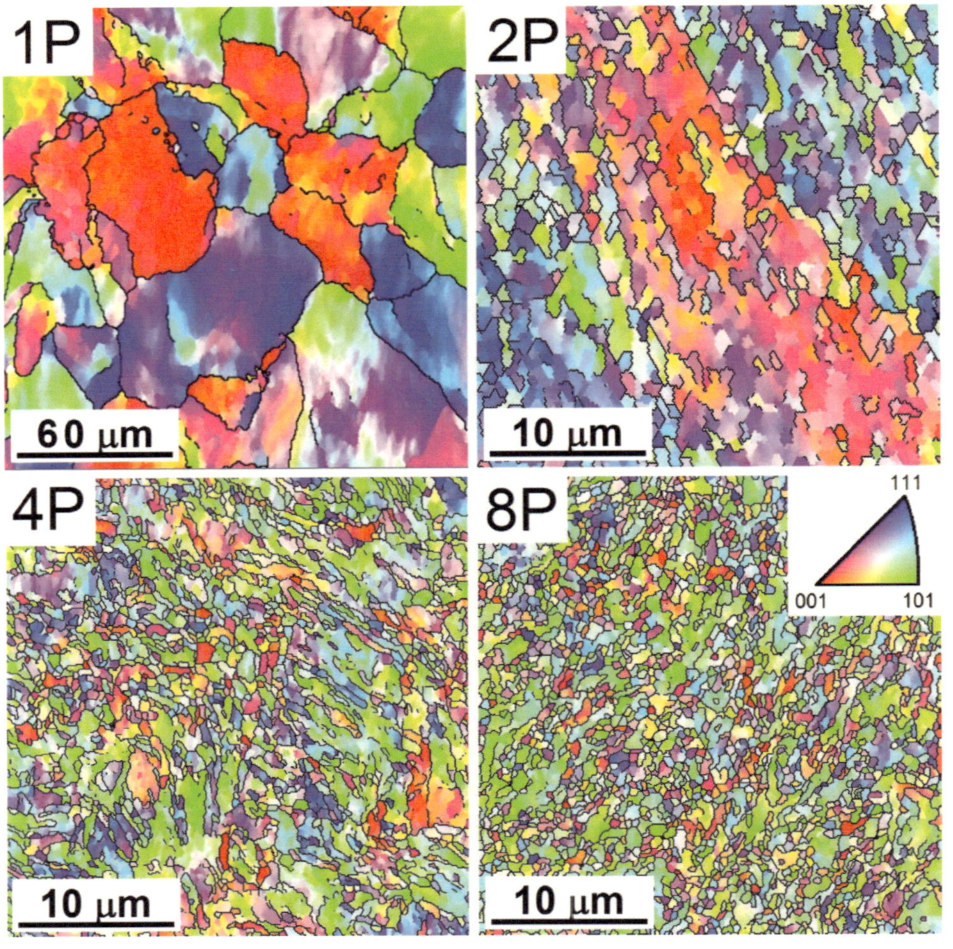

Figure 11. EBSD inverse pole figure maps obtained on the cross-sections of the samples processed by ECAP for different numbers of passes.

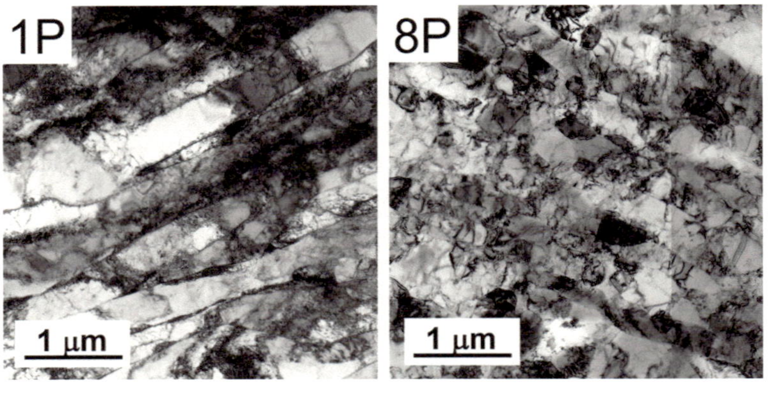

Figure 12. A bright-field TEM micrographs of IF steel samples deformed by applying (a) 1 pass and (b) 8 passes.

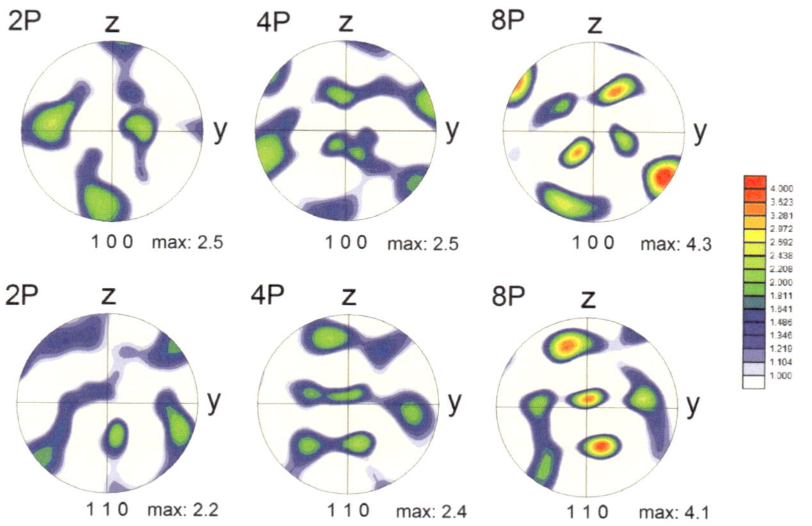

Figure 13. EBSD (100) and (110) pole figures for the IF steel samples processed by ECAP for different numbers of passes.

detailed LT investigation of deformed Fe and reported that the lifetimes of positrons trapped at edge and screw dislocations are τ_{edge} = 165 ps and τ_{screw} = 142 ps, respectively. As a consequence, the lifetime $\tau_2 \approx 150$ ps determined in IF steel deformed by ECAP can be attributed to positrons trapped at a mixture of edge and screw dislocations. The fraction of screw dislocations f_{screw} was determined from the lifetime τ_2 according to the relation:

$$f_{screw} \approx \frac{\tau_{edge} - \tau_2}{\tau_{edge} - \tau_{screw}} \tag{2}$$

The IF steel contains $f_{screw} \approx 0.7$ after the first ECAP pass. During further ECAP processing, the screw/edge character of dislocations remains almost unchanged, as indicated by almost constant value of the lifetime τ_2 in **Figure 14(a)**. More screw character of dislocations was also observed in ECAP deformed Cu as reported in Section 3.2. In BCC lattice of IF steel, the screw dislocation core is dissociated into a non-planar configuration [41]. As a consequence, during

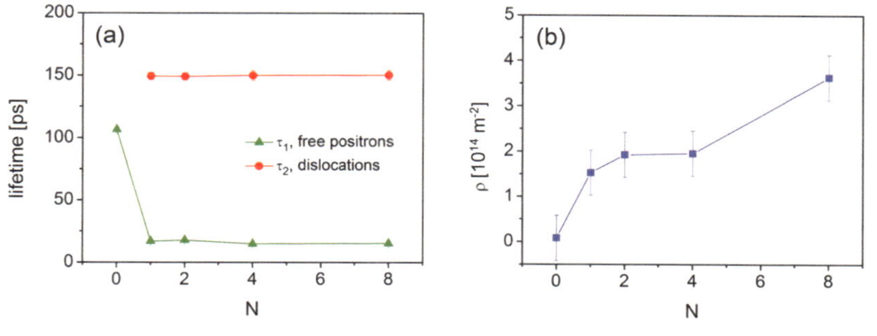

Figure 14. Development of (a) lifetimes of the components resolved in LT spectra and (b) the mean density of dislocations ρ determined by PAS with increasing number of ECAP passes. The initial sample before ECAP processing is labelled as 0N.

ECAP pressing, edge dislocation segments annihilate more easily than the screw ones and the remaining dislocations have therefore a more screw character. The development of the mean dislocation density (ρ) determined from LT data applying the DTM is plotted as a function of N in **Figure 14(b)**. The mean dislocation density gradually increases with increasing number of ECAP passes and saturates at $\approx 4 \times 10^{14}\ m^{-2}$.

4.4. Mechanical properties

The results of mechanical testing at RT showed that ECAP processing significantly influences the mechanical properties of IF steel (see **Figure 15**). The values of microhardness (*HV*) and tensile strength (σ_{max}) gradually increase with increasing N and are the same within the experimental error (see **Figure 15a**). Assuming that microhardness test introduces the multiaxial loading to the material in comparison to the uniaxial tensile test, the consistency in the values of *HV* and σ_{max} indicates a negligible effect of the texture on the mechanical properties. On the other hand, the yield stress ($\sigma_{0.2}$) significantly increases already after a single pass as a result of the rapid increase of dislocation density and simultaneous grain refinement. Further straining leads to the only moderate increase of $\sigma_{0.2}$ due to the slight increase of dislocation density and continuous grain refinement. A similar behaviour was observed in ECAP-processed aluminium alloy 6016 [42]. The overall increase in the tensile strength and the yield stress was about 230 and 450%, respectively. As it is apparent from **Figure 15(b)**, increasing strength of the samples is accompanied by a rapid decrease of ductility, which drops from 45% for initial sample to approximately 10% after eight passes. Such reduction in the ductility is typical for ECAP-processed materials and can be explained by reducing mobility of dislocations due to the considerable increase of dislocation density [43, 44]. It can be concluded that mechanical properties of ECAP processed IF steel are controlled mostly by grain size and dislocation density.

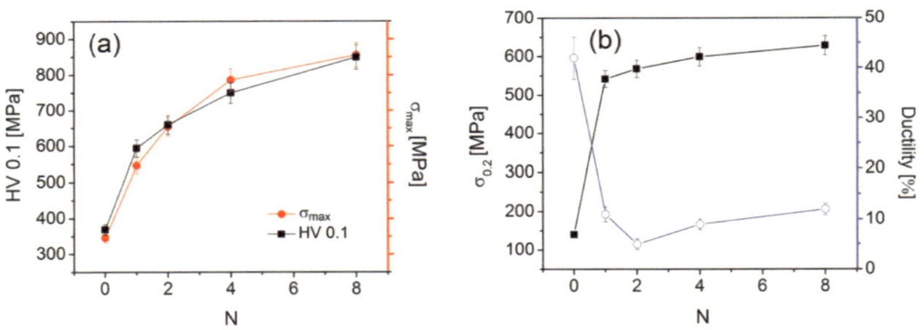

Figure 15. Development of (a) microhardness, tensile strength and (b) yield stress, ductility of IF steel with increasing numbers of ECAP passes.

5. Materials with HCP structure

Since the first introduction of ECAP, magnesium is the most intensively investigated HCP metal. Its processing is much more difficult than the processing of FCC and BCC metals, because of the limited number of available slip systems in the HCP structure at RT. Therefore, much higher processing temperature is often needed, which significantly complicates achieving average grain size below 1 μm because of grain growth at elevated temperatures. Nevertheless, the improvement in the processing technique, especially by the utilization of back pressure in the exit channel enables to decrease the processing temperature and finally to achieve much finer microstructure. In this section, the effect of the processing parameters and the composition of the alloy on the microstructure development and resulting mechanical properties in magnesium alloys are introduced and discussed.

The most important parameters that could be varied in ECAP processing are the processing temperature and the processing route. Both parameters have significant influence on the resulting microstructure, and consequently on physical properties of the final material. The processing temperature is an experimental parameter and needs to be optimized for each alloy separately. The processing temperature is usually around 200°C, but for certain alloys, particularly those with the high content of rare earth elements, the processing temperature up to 350°C needs to be used. In the next paragraphs, the effect of processing route and processing temperature selection on the example of a commercial AX41 (Mg, 4 wt%; Al, 1 wt% Ca) magnesium alloy is analysed.

5.1. Effect of the processing route

The microstructure of the extruded (EX) sample is shown in **Figure 16**. Homogeneous distribution of equiaxed grains with an average grain size of 10 μm was observed in both section planes. Texture of the extruded sample was typical for most magnesium alloys, namely, $\langle 10\bar{1}0 \rangle$ fibre texture with a fibre parallel to the extrusion direction (see **Figure 17**).

The extruded samples were subsequently processed by ECAP following three different processing routes—A, B_C and C. Eight passes through ECAP resulted in gradual refinement of the microstructure. Microstructure of specimens processed by 8P irrespective of the processing route is shown in **Figure 18(a–c)**. Nevertheless, the fragmentation rate and the final grain size depend strongly on the individual ECAP routes. The evolution of the average grain size for all samples/routes is shown in **Figure 18(d)**. The mean grain size was determined from the EBSD images as the area-weighted mean grain size. From the results, it may be concluded that the routes C and B_C were more effective during the first steps of the processing and homogeneous

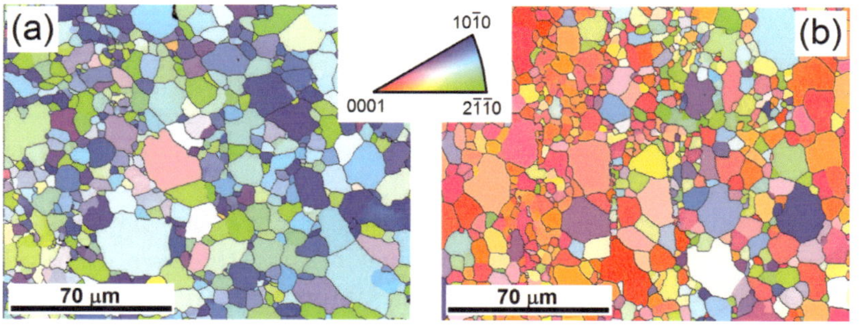

Figure 16. EBSD inverse pole figure maps obtained on the (a) cross-section and (b) longitudinal section of the extruded sample.

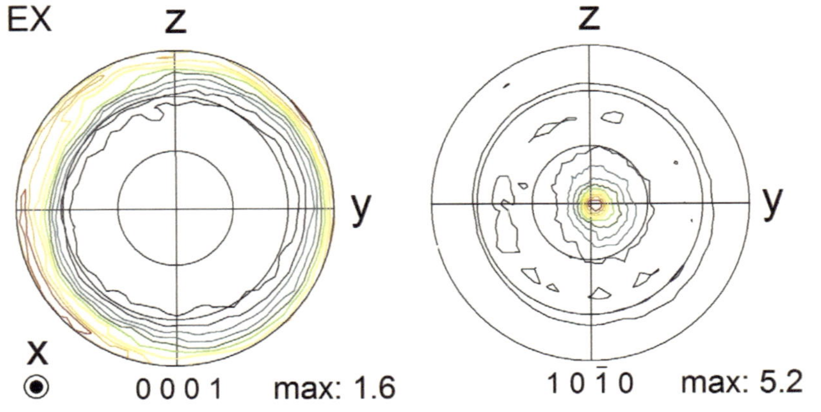

Figure 17. (0001) and (10$\bar{1}$0) X-ray pole figure measured on cross-section.

fine-grained microstructure was attained already after four passes (see Ref. [45]). The average grain size of samples processed via route B_C was unchanged during subsequent passes, but extensive grain growth was observed in samples processed via route C. The resulting value of average grin size was ~2.7 μm for route B_C and ~4.5 μm for route C. In the case of route A, the grain refinement was continuous and homogeneous microstructure was attained only after eight passes through ECAP, with the average grain size of ~2 μm.

The grain refinement during ECAP is based on two mechanisms that are working cooperatively. The first one is a nucleation and growth of fine grains along former grain boundaries and the second one is the formation of high-angle grain boundaries from dislocation tangles [46]. The first one is a more intensive refinement process in HCP structures, whereas as mentioned earlier, the second one is more intensive in FCC and BCC structures. These two mechanisms usually lead to gradual grain refinement until grain growth and grain refinement are in balance, as it was observed in the case of route B_C after 4P. Nevertheless, a different evolution of the structure and grain refinement was observed for samples processed via routes A and C. This difference could be explained by the analysis of the Burges vectors population in the individual samples. The analysis of the distribution of dislocations in the non-basal <a> slip systems, which was performed using the procedure described in detail in [9], is shown in **Figure 19**. Prismatic and pyramidal (PrE + PyE) <a>-type dislocations evolution as a function of the number of ECAP passes for different processing routes, are shown together because of the analysis limitation. The complete analysis of all major slip systems is described in detail elsewhere [45]. The significant dislocation activity observed in all samples/routes is consistent with theoretical calculations [47, 48], where modelling results indicating that approximately 20% of strain accommodated by prismatic <a>-slip are presented. The non-basal <a>-dislocations are very important for grain refinement, because they have the high probability to lock each other and form dislocation tangles even in small grains. Therefore, substantially higher fraction of these dislocations in samples processed via route A are responsible for higher grain refinement and vice versa, i.e. the reduction of non-basal <a>-dislocations fraction in samples processed via route C resulted in grain growth. The activity of a particular slip system is highly dependent on the grain orientation and therefore on the texture of the material.

http://dx.doi.org/10.5772/intechopen.68965

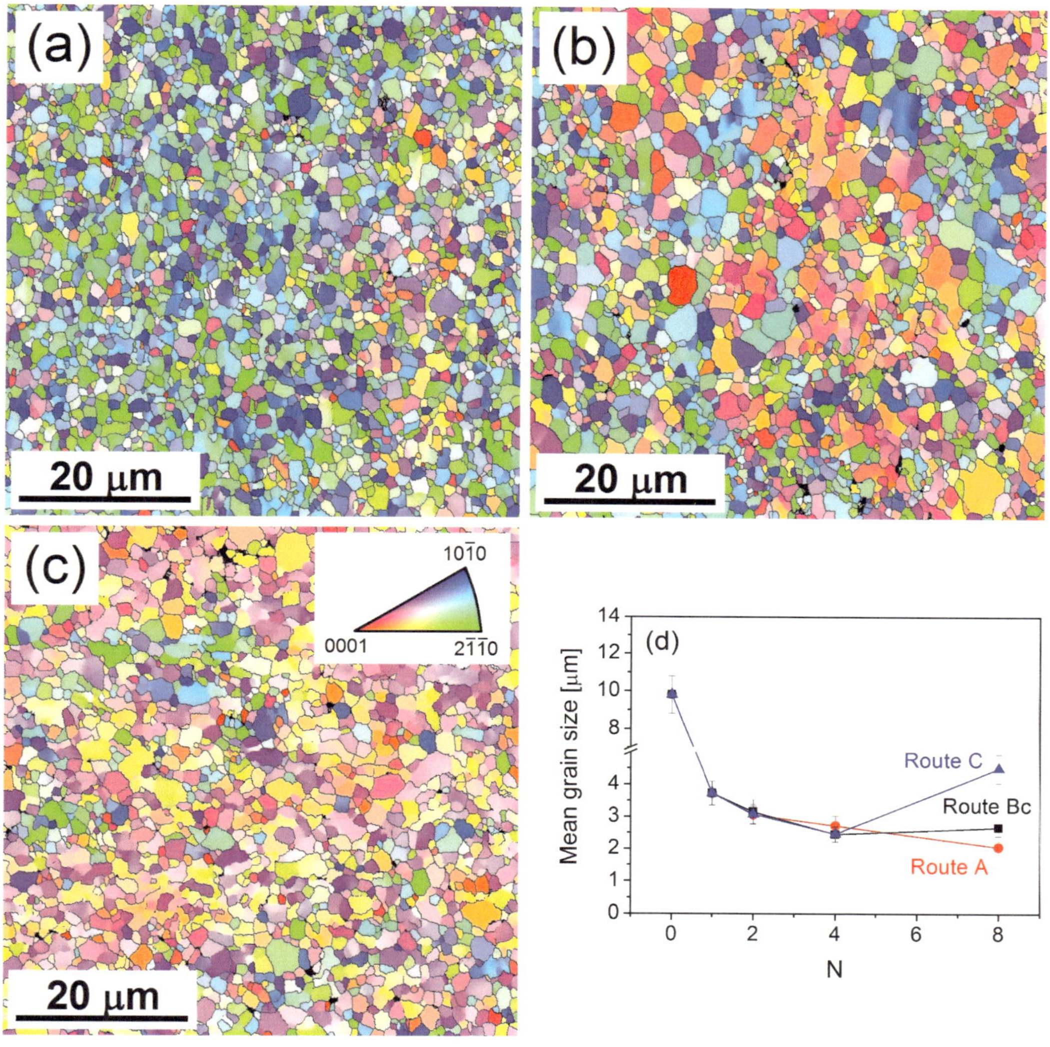

Figure 18. EBSD inverse pole figure maps obtained on the samples processed by 8 passes via route (a) A, (b) B_C and (c) C. (d) Evolution of the mean grain size in the samples processed by different routes.

Different processing routes influence significantly the texture development. Therefore, the higher activity of non-basal <a>-dislocations in samples processed via route A than route C is caused by the preferred orientation of individual grains. In **Figure 20**, pole figures measured on the samples after the final stage of the processing (8P) are displayed. Two kinds of texture components could be recognized. The first texture component, denoted as M, represents basal planes normal to the z-direction. The second texture component, denoted as N, represents basal planes lying parallel to the theoretical shearing plane activated during ECAP [49], i.e. the basal planes are tilted by ~45° to the pressing x-direction. The formation and intensity of the particular texture component is strongly influenced by the processing route. Whereas texture component M is dominant in samples processed via route A, it is completely missing in

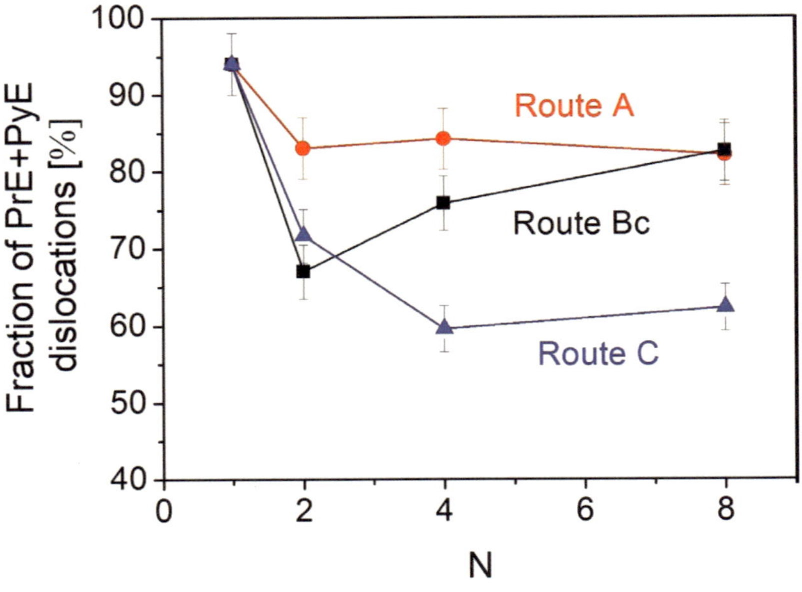

Figure 19. The dependence of the sum of relative fractions of prismatic (PrE) and pyramidal (PyE) <a>-type edge dislocations on the route of ECAP and the number of passes.

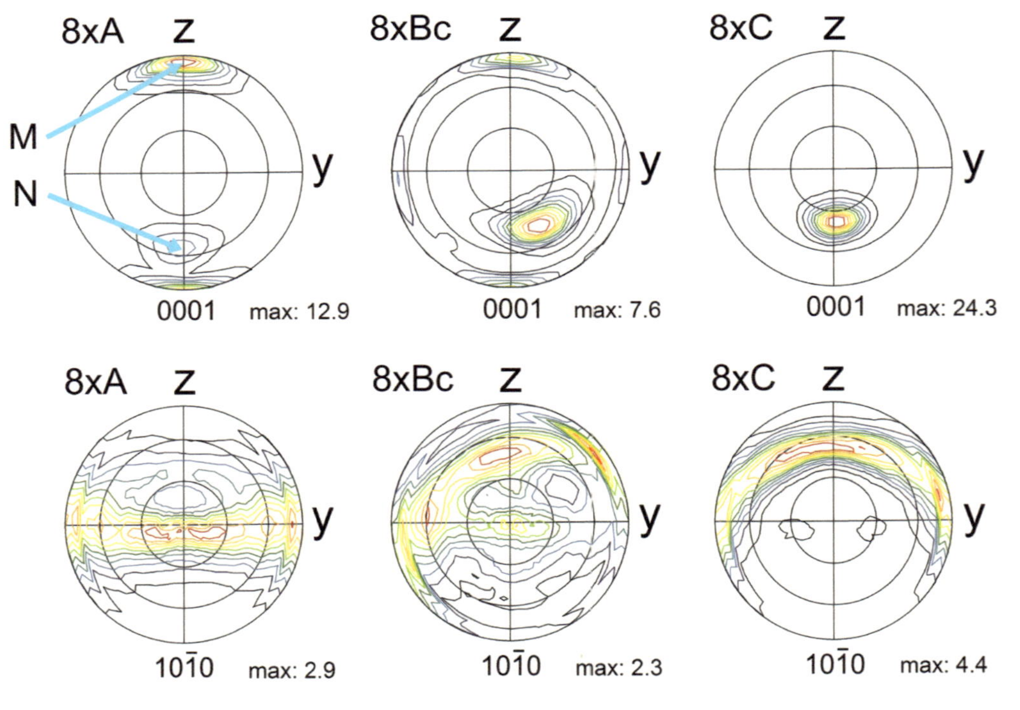

Figure 20. (0001) and $(10\bar{1}0)$ X-ray pole figures for the samples processed by different routes.

samples processed via route C. Thus, the texture component N is the only component in samples processed via route C. In the sample processed via route A, the component N is strongly suppressed. In the case of the route B_C, the situation is more complicated. The microstructure exhibits both texture components. The dominant texture component is N, which is additionally tilted roughly by 40° towards y-direction. Gradual transformation of the texture components during successive ECAP passes is described in detail elsewhere [45].

The formation mechanisms of these texture components at an expense of the initial fibre texture are the following. The texture component N is formed by predominant activation of the basal slip system during the processing, which causes the rotation of the (0001) basal planes parallel to the theoretical shearing plane (see **Figure 21b**). The formation mechanism of this texture component is discussed in different papers with the same conclusion [49–54]. The origin of the texture component M is the combination of twinning which occurs already in the feed-in channel during pressing and the activation of the second-order pyramidal slip in which the pyramidal plains $\{11\bar{2}2\}$ remain orientated parallel to ECAP shearing plane (see **Figure 21a**). For full description of the formation mechanism of both texture elements, the reader is referred to Refs. [45, 55]. The occurrence of compression twining in the extruded pure magnesium and its alloys was also reported by many authors [56, 57].

The formation of a particular texture component and its strength is given by the route of ECAP processing. Rotation of the sample after the *n*th ECAP pass, which differentiates the type of the route, defines the orientation of the shearing plane towards the pre-existing texture components during the subsequent pass. Processing of the sample without any rotation (route A) causes the generation and strengthening of the texture component M. Reorientation of individual grains towards the M component after each pass results in unfavourable orientation for the basal slip in the subsequent pass. Therefore, this texture component is very weak. Additionally, grains representing the M component are oriented well for twinning in the feed-in channel, and subsequently, the second-order pyramidal slip is activated and strengthens the texture component M, as shown in **Figure 21**.

In the case of route C, the sample is rotated 180° along its processing direction after each pass. Grains representing the texture component N have basal planes aligned with the ECAP

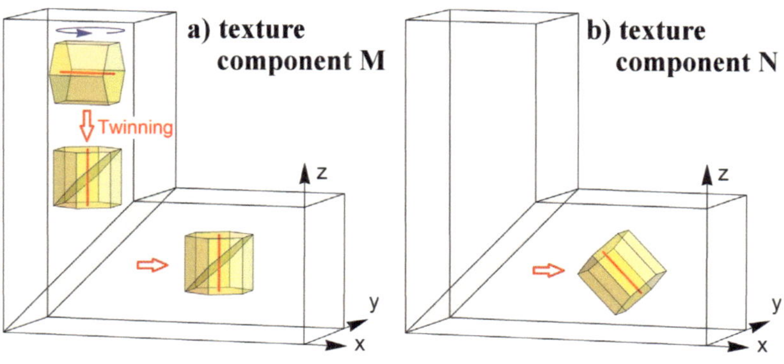

Figure 21. Formation mechanism of the texture components (a) M and (b) N.

shearing plane and therefore there is no rotation of these grains during the subsequent pass. Moreover, the activation of the basal slip in other grains results in strengthening of the component N. The formation of this texture component is very effective, and one can notice that the texture strength in the sample processed via this route is significantly higher than in samples processed via the other routes.

The pole figure of the sample processed by route B_C is a combination of the previous two ones. Rotation of the sample by 90° along the processing direction partially suppresses the strengthening mechanism of the N component. The initial extrusion texture is favourable for the formation of the M component, because the majority of grains are well oriented for twinning in the feed-in channel. However, the fraction of these grains gradually decreases as more grains are reoriented to form the N component. In this orientation, the grains can no longer twin in the feed-in channel and consequently generate the M component. Therefore, in some works describing the evolution of the texture in Mg alloys processed via route B_C, the texture component M is present after the final pass and in some is not. Its presence is given by the effectivity of twinning of the grains in the feed-in channel and the rate by which the texture component N strengthens. This rate is affected particularly by the processing temperature, as discussed below.

5.2. Effect of the processing temperature

The effect of the processing temperature is more straightforward than the effect of the other processing parameters. The higher processing temperature makes usually the processing itself easier while it increases the tendency for the grain growth. Magnesium must be processed at elevated temperatures because of the billet segmentation, which occurs at the lower temperatures [58]. The high limit for the processing temperature fundamentally does not exist, but it is always necessary to process at the lowest possible temperature to obtain the most effective grain refinement. In the previous section, the grain refinement and texture formation for the extruded AX41 processed at 220°C via route B_C were discussed. This temperature was found to be the lowest one for this alloy. In this section, the effect of increase of the processing temperature to 250°C is shown and discussed.

The microstructure of the sample processed at 250°C temperature was comparable to that processed at 220°C. Nevertheless, the apparent negative effect of the increased processing temperature on the grain size is clearly seen in **Figure 22**. At 250°C, the grain refinement was observed only up to two passes. During further straining ($N > 2$), the grain growth occurs. The final average grain size was ~4 μm, while for lower processing temperature (220°C), it was ~2.7 μm only. Another significant effect of the increase of the processing temperature was observed in texture development. In **Figure 23**, pole figures of the 8P sample processed at 250°C are shown. The formation of the texture component M is significantly suppressed at the higher temperature, cf. **Figure 20**. This difference results from the suppression of twinning at elevated temperatures. As a result, after eight passes, the M component is strongly suppressed in specimen deformed at the higher temperature.

5.3. Effect of the composition

The influence of the composition on the deformation behaviour and resulting microstructure evolution in magnesium alloys has been intensively studied for a variety of processing

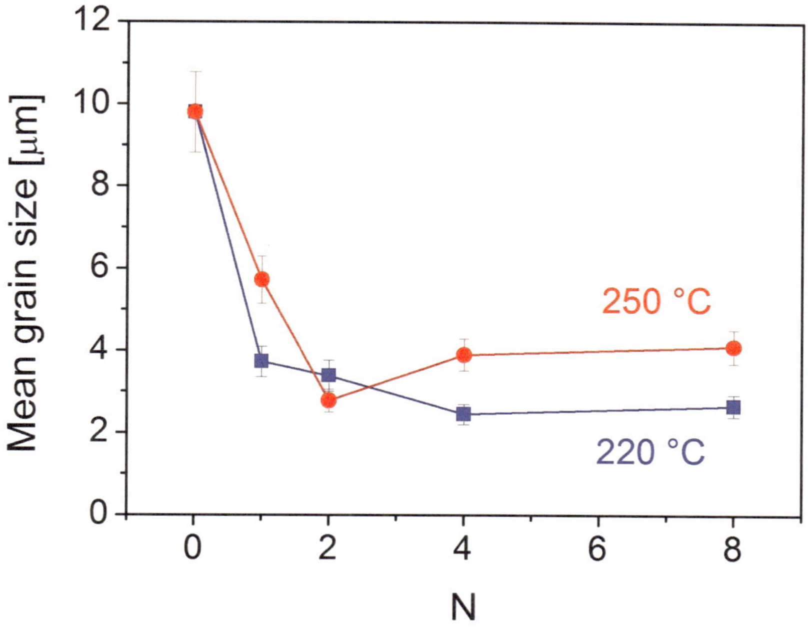

Figure 22. Evolution of the mean grain size on the processing temperature and number of ECAP passes.

techniques, especially for extrusion and rolling. Nevertheless, in the case of ECAP, the usual selection of alloys is usually limited to AZ, AM, and ZK types of alloys. The alloying elements in these types of alloys usually form stable secondary phases, or only small quantities of the atoms are dissolved in the Mg matrix. Therefore, the effect of the alloying elements on the deformation behaviour of the Mg matrix is highly limited. On the other hand, strong effect was observed in magnesium alloys containing lithium.

The effect of lithium on the final microstructure formed by ECAP in extruded AE42 (Mg, 4 wt%; Al, 2 wt% rare earths) and LAE442 (Mg, 4 wt%; Li, 4 wt%; Al, 2 wt% rare earths) magnesium alloys will be discussed in this section. The only difference between these two alloys is in the

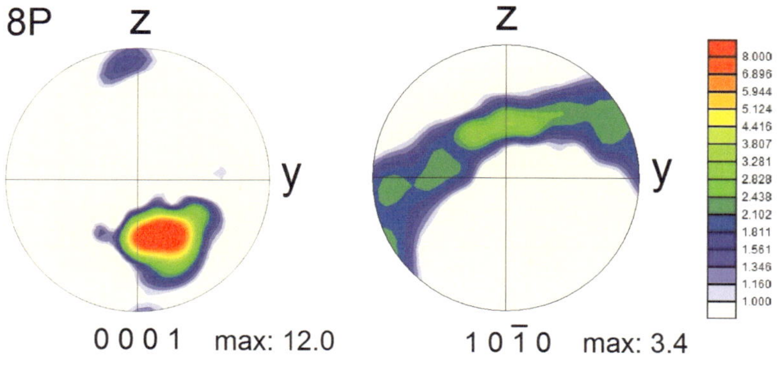

Figure 23. (0001) EBSD pole figure of the sample processed at 250°C (X-plane).

presence of lithium in the latter one. Lithium content in the alloy is below the solubility limit. It is very important as exceeding the solubility limit results in the formation of the mixture of HCP and BCC structure [59]. XRD measurement focused on the investigation of the effect of Li content on the lattice parameters of this alloy and showed a decrease of the *c/a* ratio to 1.610 [46], when compared to the pure Mg with *c/a* = 1.624 [60]. The decrease of the *c/a* ratio has a strong effect on the deformation behaviour as it facilitates the activation of non-basal slip systems.

The microstructure of both alloys was investigated by EBSD. Both alloys were processed by ECAP in the temperature range 180–220°C following route B_C up to 8 and 12P for AE42 and LAE442 alloys, respectively. The corresponding micrographs are shown in **Figure 24**. After the final step ECAP, the homogeneous microstructure with comparable grain size in both alloys of about ~1.5 μm was achieved. Effect of ECAP on grain refinement observed in both alloys is detailed elsewhere [55]. It should be noted that the reason for the processing by a higher number of passes of the LAE442 alloy was a much higher grain size in the initial condition. The effect of the grain size in the initial condition of the processed material on the effectivity of grain refinement was discussed in [61]. Our results are fully consistent with this work.

The microstructure of both alloys looks very similar. However, there is a significant difference in the texture, which has developed during ECAP processing. The corresponding pole figures are shown in **Figure 25**. The standard and usual texture was observed in both alloys after the extrusion [55]. The pole figure of the processed AE42 alloy is very similar to the texture observed in the AX41 alloy processed by ECAP (cf. **Figure 20**). On the other hand, the pole figure determined in the processed LAE442 alloy is completely different, even if the same processing conditions were employed. The pole figure contains a very strong M component, a weak N component and a new third component with basal planes perpendicular to the y-direction. The formation mechanism of the M and N components resulting from the predominant activation

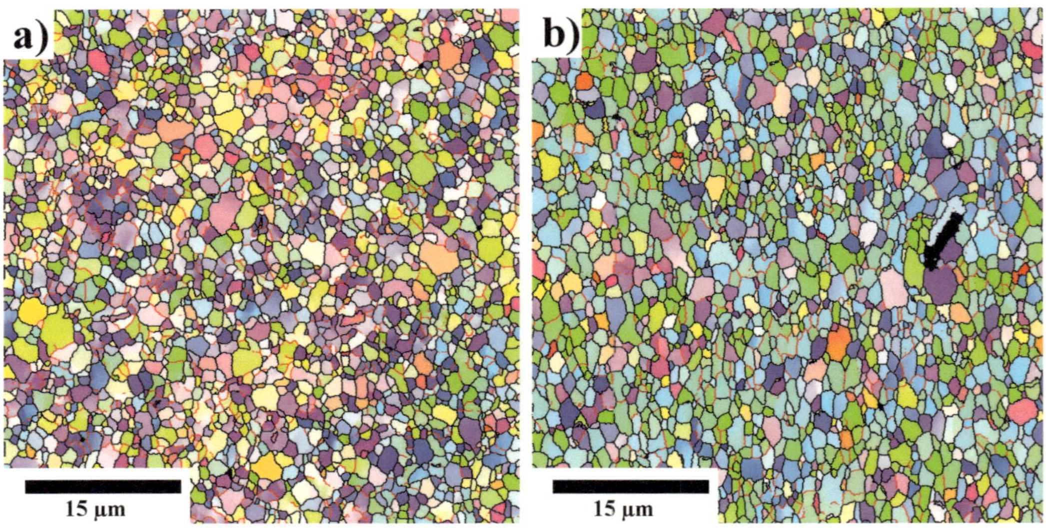

Figure 24. EBSD inverse pole figure maps of (a) AE42 and (b) LAE442 after the final step of ECAP.

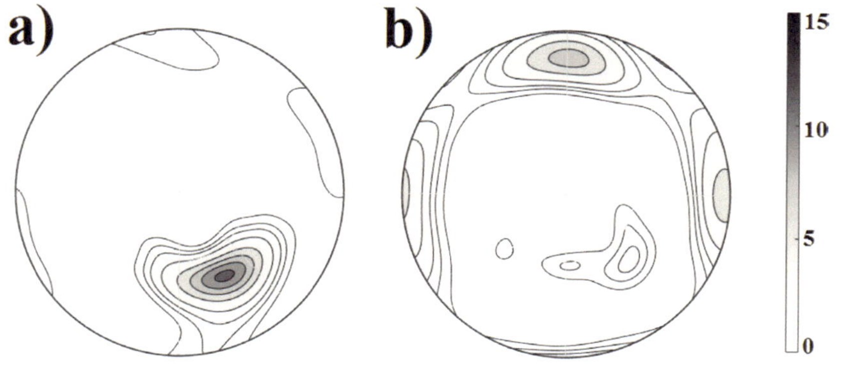

Figure 25. (0001) X-ray pole figure of (a) AE42 and (b) LAE442 after the final step of ECAP.

of the second-order pyramidal and basal slip, respectively, were discussed in Section 5.1. The new texture element, denoted as L, is associated with the activity of the prismatic slip system from the analysis of the $(10\bar{1}0)$ pole figure (see **Figure 26**). The main maximum is present on the z-axis and rotated by 15° from the processing direction. As described above and in Ref. [52], the rotation of the active slip planes parallel to the shear plane of the ECAP die causes the formation of the other texture element. Similarly, activation of the $\{(10\bar{1}0)\}\langle 11\bar{2}0\rangle$ prismatic slip system due to the lithium addition caused the rotation of the prismatic planes parallel to the shear plane of ECAP die as graphically expressed in **Figure 27**. Geometrically, this rotation causes the formation of the texture elements presented schematically in **Figure 26**.

5.4. Mechanical properties

In the previous section, it was shown that the microstructure of the processed magnesium alloy may be very different depending on the various processing parameters and the composition. Mechanical properties of the material are strongly affected by the microstructure, and therefore different evolution of the mechanical properties is expected. There are three

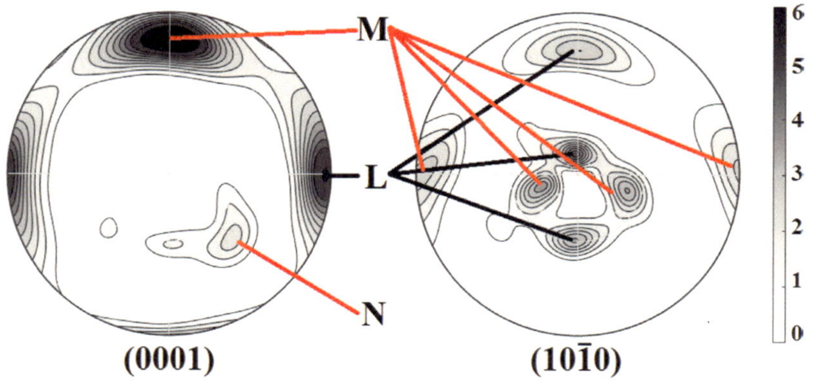

Figure 26. Depiction of the texture elements in the processed LAE442 alloy.

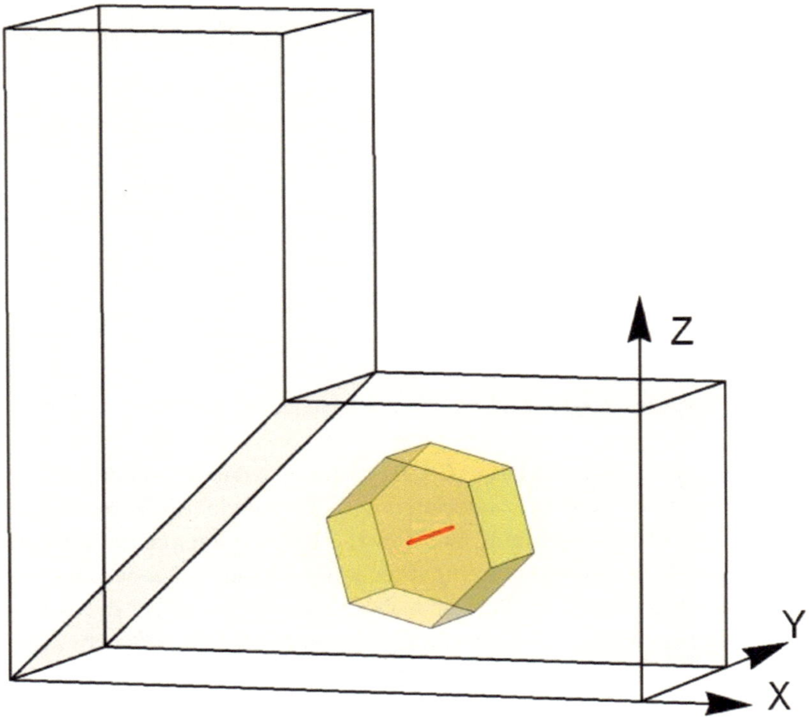

Figure 27. The formation mechanism of the texture component L.

main factors affecting the mechanical strength, namely, the grain size, the texture and the dislocation density. It is very hard to separate the effect of these individual parameters. However, the effect of the texture can be strongly suppressed by microhardness measurement, which simulates the multiaxial-loading. Therefore, it gives the opportunity to reveal the effect of grain refinement and hardening through dislocations. In **Figure 28(a)**, the dependence of the microhardness on the number of ECAP passes is shown. The microhardenss in individual samples increases with increasing number of ECAP passes and then saturates. The only exception is the AX41 sample processed via route C, in which the grain growth was observed. Grain boundary hardening is therefore an obvious source of the enhanced microhardness. Nevertheless, the possible effect of dislocations may also be assessed, when the data are evaluated considering the Hall-Petch relation. The highest difference in the discussed alloys was observed between AE42 and LAE442. **Figure 28(b)** shows the dependence of the microhardness on the square root of grain size for both alloys. The samples of the AE42 alloy obey the linear tendency of Hall-Petch relation, while in the LAE442 alloy, Hall-Petch relation is not met. The non-linear tendency and higher values of microhardness are the result of the increased dislocation density. **Figure 28(c)** shows the evolution of the dislocation density measured by PAS for both alloys. The evolution is very similar, but the values measured in the LAE442 alloy are by one order of magnitude higher than in the AE42 alloy. As a consequence, the effect of dislocations on the mechanical strength needs to be added to the calculation.

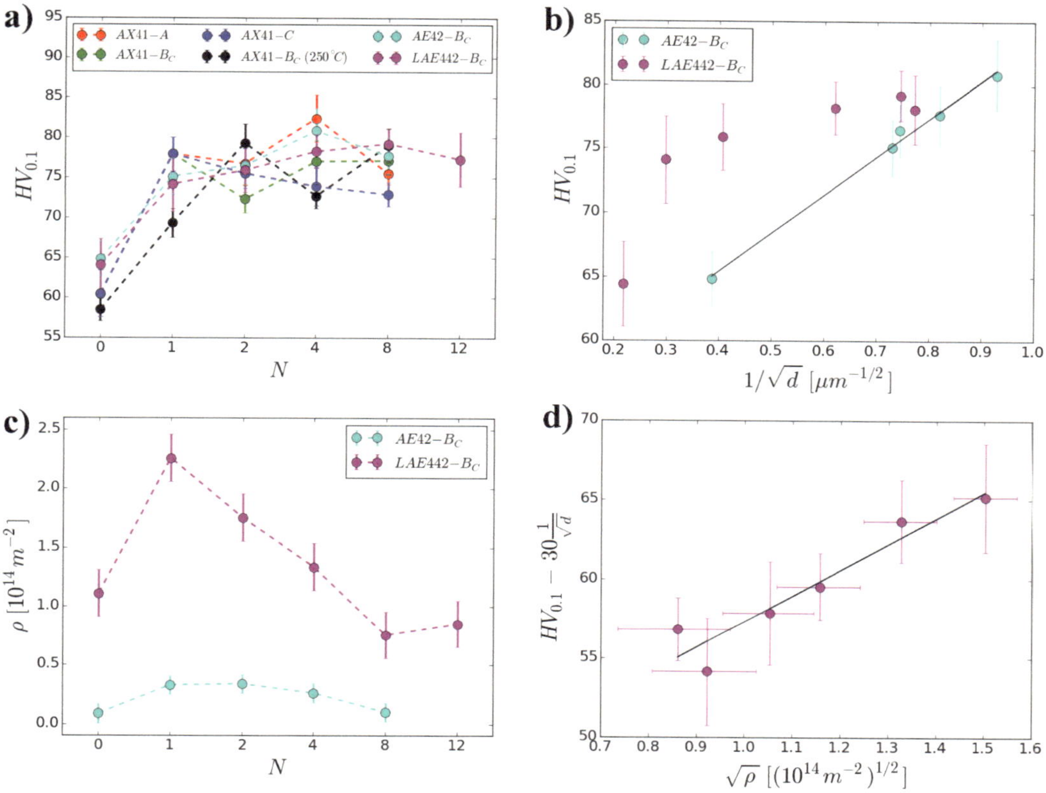

Figure 28. Evolution of microhardness in investigated Mg alloys. (a) The evolution of microhardnes different Mg alloys, (b) Hall-Petch relation for the AE42 and LAE442 alloys, (c) evolution of dislocation density in the AE42 and LAE442 alloys, (d) the evolution of reduced microhardness by grain boundary hardening as a function of square root of dislocation denisty in the LAE442 alloy.

Assuming that the grain boundary hardening is similar in both alloys (hardening coefficient was 30), it can be subtracted from the microhardenss and net effect of dislocations could be revealed. As shown in **Figure 28(d)**, the *HV* values corrected for the grain boundary hardening obey the linear relation when plotted vs. square root of the dislocation density. Variations of HV values of the AX41 alloy in **Figure 28(a)** are caused by variations of dislocation density. A detailed discussion of this effect is given elsewhere [45].

The evolution of the yield tensile strength shown in **Figure 29(a)** differs significantly from the evolution of the microhardness. The difference is caused by the mode of loading, i.e. by the change from multiaxial (microhardness) to an uniaxial loading (tensile tests), in which texture plays a significant role. The positive effect of the grain refinement and increased dislocation density on the mechanical stength can be overwhelmed by a negative effect of the texture that developes during ECAP. As shown above, strong texture is formed in investigated alloys regardeless of the processing route. Therefore, the strong anisotropy of the mechanical properties of the single crystal is transfered to the final material. The ECAP billet has ususally a form of a rod or a bar, and therefore the mechanical properties are ususally investigated in

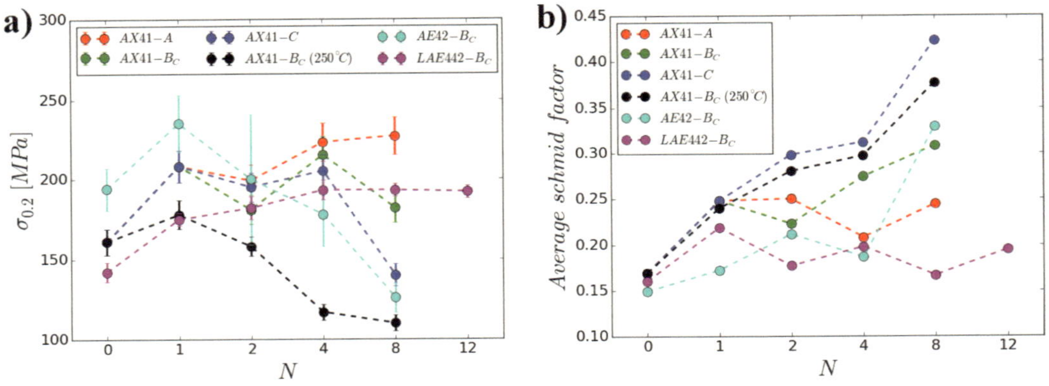

Figure 29. Mechanical properties of investigated Mg alloys. (a) The evolution of yield tenisle stress with strain, (b) the evolution of average Schmid factor calculated for uniaxial tension paralel to X-axis from EBSD data.

the direction paralel to the processing direction. Different evolution of mechanical proeprties was observed in individual samples even if the microstructure is strongly refined in all cases. As mentioned above, the reason is a strong texture formed in all samples, which differs in individual processing routes. The measure of the texture effect is the Schmid factor calculated for the basal slip (m_{basal}). Basal slip is the most favourable slip at room temperature, because its critical resolve shear stress is much lower than that of other slip systems. The plot of m_{basal} as a function of increasing number of ECAP pases is shown in **Figure 29(b)**. The strong correlation between m_{basal} and the tensile yield stress is clearly seen (cf. **Figure 29a** and **b**). The samples with higher value of m_{basal} exhibit lower yield stress. Deterioration of the yield tensile stress along the procesing direction is caused by the formation of the texture component N. Gradual reorientation of the grains by ~45° from the processing direction and also from the deformation axis facilitates the activation of the basal slip system, as manifested by enhanced values of m_{basal}. On the other hand, the formation of the strong texture component M in specimens processed by route A results in the same deterioration of mechanical properties when the deformation axis is rotated. As a consequence, the use of route A does not suppress the negative texture effect. The lowest anisotropy of mechanical properties was observed in the LAE442 alloy, where the final texture was not so sharp and all three texture components were present.

6. Conclusions

Microstructure and lattice defect evolution in ultrafine-grained materials of different crystal structure processed by ECAP have been investigated as a function of strain imposed to the material by severe plastic deformation and correlated with mechanical properties of these materials. The following conclusions may be drawn from this thorough study:

- ECAP proved to be an efficient technique of grain refinement. Ultrafine-grained materials of different crystal structure are obtained by this technique.
- The extent and the rate of the grain refinement strongly depend on the amount of the imposed strain and the parameters of ECAP processing.

- ECAP is more effective in FCC and BBC materials as compared to hexagonal materials. The difference is caused by different temperatures of processing and the concomitant processes occurring during pressing. Cubic materials may be processed at room temperature while elevated temperatures are needed to process HPC materials.
- In HCP materials, the dynamic recrystallization is the most effective refinement mechanism, while in BCC and FCC materials a dominant mechanism of grain refinement is controlled by the gradual transformation of LAGBs to HAGBs.
- ECAP-processed materials exhibit strong a crystallographic texture, which depends on the chemical composition, the temperature and the processing route.
- A high density of lattice defects was found in severely deformed material processed by ECAP. Significantly higher density of dislocations was found in materials with cubic structure than in hexagonal materials (by at least one order of magnitude). FCC metals also contain a high concentration of vacancies and their conglomerates called microvoids while in BCC and hexagonal materials vacancy concentration is very low (below the detection limit of PAS).
- Positron annihilation spectroscopy and X-ray diffraction proved to be efficient and complementary techniques of the determination of the dislocation density over a wide range of orders of magnitudes.
- Processing of materials having FCC and BCC lattice leads to the enhanced strength, which is accompanied by reduced ductility.
- Improved strength of UFG materials is caused by a combination of strong grain refinement and increased dislocation density.
- Mechanical properties in HCP materials are significantly affected by the grain size, dislocation density and texture, which results in their strong anisotropy.

Acknowledgements

This work was financially supported by the Czech Science Foundation under the project GB 14-36566G and by ERDF under the project 'Nanomaterials centre for advanced applications', project No. CZ.02.1.01/0.0/0.0/15_003/0000485.

Author details

Peter Minárik[1], Tomáš Krajňák[1], Ondřej Srba[2], Jakub Čížek[1], Jenő Gubicza[3], Milan Dopita[1], Radomír Kužel[1] and Miloš Janeček[1]*

*Address all correspondence to: janecek@met.mff.cuni.cz

1 Charles University, Prague, Czech Republic

2 Research Center Rez, Czech Republic

3 L. Eotvos Unviersity, Budapest, Hungary

References

[1] Sabirov I, Enikeev NA, Murashkin MY, Valiev RZ. Bulk Nanostructured Materials with Multifunctional Properties. Berlin/Heidelberg: Springer; 2015. p. 125

[2] Iwahashi Y, Horita Z, Nemoto M, Langdon TG. Factors influencing the equilibrium grain size in equal-channel angular pressing: Role of Mg additions to aluminum. Metallurgical and Materials Transactions A. 1998;**29**(10):2503-2510

[3] Nakashima K, Horita Z, Nemoto M, Langdon TG. Development of a multi-pass facility for equal-channel angular pressing to high total strains. Materials Science and Engineering: A. 2000;**281**(1-2):82-87

[4] Raab GJ, Valiev RZ, Lowe TC, Zhu YT. Continuous processing of ultrafine grained Al by ECAP–Conform. Materials Science and Engineering: A. 2004;**382**(1-2):30-34

[5] West RN. Positron studies of lattice defects in metals. In: Hautojärvi PDP, editor. Positrons in Solids. Berlin/Heidelberg: Springer; 1979. pp. 89-144 (Topics in Current Physics)

[6] Oberdorfer B, Würschum R. Positron trapping model for point defects and grain boundaries in polycrystalline materials. Physical Review B. 2009;**79**(18):184103

[7] Čížek J, Procházka I, Cieslar M, Kužel R, Kuriplach J, Chmelík F, et al. Thermal stability of ultrafine grained copper. Physical Review B. 2002;**65**(9):094106

[8] Ribárik G, Gubicza J, Ungár T. Correlation between strength and microstructure of ball-milled Al–Mg alloys determined by X-ray diffraction. Materials Science and Engineering: A. 2004;**387-389**:343-347

[9] Máthis K, Nyilas K, Axt A, Dragomir-Cernatescu I, Ungár T, Lukáč P. The evolution of non-basal dislocations as a function of deformation temperature in pure magnesium determined by X-ray diffraction. Acta Materialia. 2004;**52**(10):2889-2894

[10] Williams DB, Carter CB. Transmission Electron Microscopy: A Textbook for Materials Science. 2nd ed. New York: Springer; 2009. p. 820

[11] Schwartz AJ, Kumar M, Field DP, Adams BL, editors. Electron Backscatter Diffraction in Materials Science. 1st ed. New York: Springer; 2000. p. 350

[12] Furukawa M, Iwahashi Y, Horita Z, Nemoto M, Langdon TG. The shearing characteristics associated with equal-channel angular pressing. Materials Science and Engineering: A. 1998;**257**(2):328-332

[13] Dalla Torre F, Lapovok R, Sandlin J, Thomson PF, Davies CHJ, Pereloma EV. Microstructures and properties of copper processed by equal channel angular extrusion for 1-16 passes. Acta Materialia. 2004;**52**(16):4819-4832

[14] Janeček M, Hadzima B, Hellmig RJ, Estrin Y. The influence of microstructure on the corrosion properties of cu polycrystals prepared by ecap. Kov Materiäly. 2005;**43**(4):258-271

[15] Dopita M, Janeček M, Rafaja D, Uhlíř J, Matěj Z, Kužel R. EBSD investigation of the grain boundary distributions in ultrafine-grained Cu and Cu–Zr polycrystals prepared by equal-channel angular pressing. International Journal of Materials Research. 2009;**100**(6):785-789

[16] Christian JW, Mahajan S. Deformation twinning. Progress in Materials Science. 1995;**39**(1):1-157

[17] Huang CX, Wang K, Wu SD, Zhang ZF, Li GY, Li SX. Deformation twinning in polycrystalline copper at room temperature and low strain rate. Acta Materialia. 2006;**54**(3):655665

[18] McKee BTA, Saimoto S, Stewart AT, Stott MJ. Positron trapping at dislocations in copper. Canadian Journal of Physics. 1974;**52**(9):759-765

[19] Barbiellini B, Puska MJ, Torsti T, Nieminen RM. Gradient correction for positron states in solids. Physical Review B. 1995;**51**(11):7341-7344

[20] Ribárik G, Ungár T, Gubicza J. MWP-fit: A program for multiple whole-profile fitting of diffraction peak profiles by ab initio theoretical functions. Journal of Applied Crystallography. 2001;**34**(5):669-676

[21] Matěj Z, Nichtová L, Kužel R. Expanding FOX for microstructural analysis. Materials and Structures. **15**(1)

[22] Gubicza J, Nam NH, Balogh L, Hellmig RJ, Stolyarov VV, Estrin Y, et al. Microstructure of severely deformed metals determined by X-ray peak profile analysis. Journal of Alloys and Compounds. 2004;**378**(1-2):248-252

[23] Caillard D. Kinetics of dislocations in pure Fe. Part I. In situ straining experiments at room temperature. Acta Materialia. 2010;**58**(9):3493-3503

[24] Ungár T, Tichy G, Gubicza J, Hellmig RJ. Correlation between subgrains and coherently scattering domains. Powder Diffraction. 2005;**20**(4):366-375

[25] Kuzel R, Cherkaska V, Matej Z, Janecek M, Cizek J, Dopita M. Structural studies of submicrocrystalline copper and copper composites by different methods. Zeitschrift für Kristallographie Supplements. 2008;**27**:73-80

[26] Máthis K, Krajnák T, Janeček M, Dopita M, Kim HS. Microstructural evolution of equal-channel angular pressed interstitial-free steel. International Journal of Materials Research. 2009;**100**(6):834-837

[27] Máthis K, Krajňák T, Kužel R, Gubicza J. Structure and mechanical behaviour of interstitial-free steel processed by equal-channel angular pressing. Journal of Alloys and Compounds. 2011;**509**(8):3522-3525

[28] Čížek J, Janeček M, Krajňák T, Stráská J, Hruška P, Gubicza J, et al. Structural characterization of ultrafine-grained interstitial-free steel prepared by severe plastic deformation. Acta Materialia. 2016;**105**:258-272

[29] Langdon TG. The principles of grain refinement in equal-channel angular pressing. Materials Science and Engineering: A. 2007;**462**(1-2):3-11

[30] Xu C, Furukawa M, Horita Z, Langdon TG. The evolution of homogeneity and grain refinement during equal-channel angular pressing: A model for grain refinement in ECAP. Materials Science and Engineering: A. 2005;**398**(1-2):66-76

[31] Valiev RZ, Langdon TG. Principles of equal-channel angular pressing as a processing tool for grain refinement. Progress in Materials Science. 2006;**51**(7):881-981

[32] Shin DH, Kim W-J, Choo WY. Grain refinement of a commercial 0.15%C steel by equal-channel angular pressing. Scripta Materialia. 1999;**41**(3):259-262

[33] Bozorth RM. The orientations of crystals in electrodeposited metals. Physical Review. 1925;**26**(3):390-400

[34] De Messemaeker J, Verlinden B, Van Humbeeck J. Texture of IF steel after equal channel angular pressing (ECAP). Acta Materialia. 2005;**53**(15):4245-4257

[35] Baczynski J, Jonas JJ. Texture development during the torsion testing of α-iron and two IF steels. Acta Materialia. 1996;**44**(11):4273-4288

[36] Toth LS, Jonas JJ, Daniel D, Bailey JA. Texture development and length changes in copper bars subjected to free end torsion. Texture Stress Microstructure. 1992;**19**(4):245-262

[37] Montheillet F, Cohen M, Jonas JJ. Axial stresses and texture development during the torsion testing of Al, Cu and α-Fe. Acta Materialia. 1984;**32**(11):2077-2089

[38] Park Y-K, Waber JT, Meshii M, Snead CL, Park CG. Dislocation studies on deformed single crystals of high-purity iron using positron annihilation: Determination of dislocation densities. Physical Review B. 1986;**34**(2):823-836

[39] Čížek J, Procházka I, Kočík J, Keilová E. Positron lifetime study of reactor pressure vessel steels. Physica Status Solidi A. 2000;**178**(2):651-662

[40] Hidalgo C, González-Doncel G, Linderoth S, San Juan J. Structure of dislocations in Al and Fe as studied by positron-annihilation spectroscopy. Physical Review B. 1992;**45**(13):7017-7021

[41] Ngan AHW, Wen M. Atomistic simulation of energetics of motion of screw dislocations in bcc Fe at finite temperatures. Computational Materials Science. 2002;**23**(1-4):139-145

[42] Mckenzie PWJ, Lapovok R. ECAP with back pressure for optimum strength and ductility in aluminium alloy 6016. Part 2: Mechanical properties and texture. Acta Materialia. 2010;**58**(9):3212-3222

[43] Kim HS, Ryu WS, Janecek M, Baik SC, Estrin Y. Effect of equal channel angular pressing on microstructure and mechanical properties of IF steel. Advanced Engineering Materials. 2005;**7**(1-2):43-46

[44] Duan Y, Tang L, Xu G, Deng Y, Yin Z. Microstructure and mechanical properties of 7005 aluminum alloy processed by room temperature ECAP and subsequent annealing. Journal of Alloys and Compounds. 2016;**664**:518-529

[45] Krajňák T, Minárik P, Gubicza J, Máthis K, Kužel R, Janeček M. Influence of equal channel angular pressing routes on texture, microstructure and mechanical properties of extruded AX41 magnesium alloy. Materials Characterization. 2017;**123**:282-293

[46] Minárik P, Král R, Pešička J, Daniš S, Janeček M. Microstructure characterization of LAE442 magnesium alloy processed by extrusion and ECAP. Materials Characterization. 2016;**112**:1-10

[47] Li S. Orientation stability in equal channel angular extrusion. Part II: Hexagonal close-packed materials. Acta Materialia. 2008;**56**(5):1031-1043

[48] Agnew SR, Mehrotra P, Lillo TM, Stoica GM, Liaw PK. Texture evolution of five wrought magnesium alloys during route A equal channel angular extrusion: Experiments and simulations. Acta Materialia. 2005;**53**(11):3135-3146

[49] Lin HK, Huang JC, Langdon TG. Relationship between texture and low temperature superplasticity in an extruded AZ31 Mg alloy processed by ECAP. Materials Science and Engineering: A. 2005;**402**(1-2):250-257

[50] Janeček M, Yi S, Král R, Vrátná J, Kainer KU. Texture and microstructure evolution in ultrafine-grained AZ31 processed by EX-ECAP. Journal of Materials Science. 2010;**45**(17):4665-4671

[51] Kim WJ, An CW, Kim YS, Hong SI. Mechanical properties and microstructures of an AZ61 Mg Alloy produced by equal channel angular pressing. Scripta Materialia. 2002;**47**(1):39-44

[52] Agnew SR, Horton JA, Lillo TM, Brown DW. Enhanced ductility in strongly textured magnesium produced by equal channel angular processing. Scripta Materialia. 2004;**50**(3):377-381

[53] Kim WJ, Hong SI, Kim YS, Min SH, Jeong HT, Lee JD. Texture development and its effect on mechanical properties of an AZ61 Mg alloy fabricated by equal channel angular pressing. Acta Materialia. 2003;**51**(11):3293-3307

[54] Liu T, Wang YD, Wu SD, Lin Peng R, Huang CX, Jiang CB, et al. Textures and mechanical behavior of Mg–3.3%Li alloy after ECAP. Scripta Materialia. 2004;**51**(11):1057-1061

[55] Minárik P, Král R, Čížek J, Chmelík F. Effect of different c/a ratio on the microstructure and mechanical properties in magnesium alloys processed by ECAP. Acta Materialia. 2016;**107**:83-95

[56] Yin SM, Wang CH, Diao YD, Wu SD, Li SX. Influence of grain size and texture on the yield asymmetry of Mg-3Al-1Zn alloy. Journal of Materials Science & Technology. 2011;**27**(1):29-34

[57] E. Sukedai TY. Investigation of tensile–compressive yield asymmetry and the role of deformation twin in extruded pure magnesium. International Journal of Materials Research. 2010;**101**(6):736-740

[58] Langdon TG. Twenty-five years of ultrafine-grained materials: Achieving exceptional properties through grain refinement. Acta Materialia. 2013;**61**(19):70357059

[59] Massalski TB, Okamoto H. Binary Alloy Phase Diagrams. 2nd ed. Materials Park, OH: The Materials Information Society; 1990

[60] Pekguleryuz M, Kainer K, Kaya A. Fundamentals of Magnesium Alloy Metallurgy. Elsevier; 2013. p. 381

[61] Figueiredo RB, Langdon TG. Principles of grain refinement in magnesium alloys processed by equal-channel angular pressing. Journal of Materials Science. 2009;**44**(17): 4758-4762

Section 2

SPD: Modeling

Chapter 3

Numerical and Experimental Study on Constrained Groove Pressing

Yanjin Guan and Zongshen Wang

Additional information is available at the end of the chapter

http://dx.doi.org/10.5772/intechopen.68504

Abstract

Constrained groove pressing (CGP) is a new severe plastic deformation method suitable for producing ultra-fine grained sheet metals. Based on Taguchi optimization method, the influence of processing parameters such as groove width, groove angle, friction coefficient and deformation rate on deformation homogeneity of constrained groove pressing (CGP) was studied numerically utilizing DEFORM-3D. A multi-pass CGP was carried out on 1060 commercially pure aluminium, copper and Ni sheets. Through a series of experimental research, the evolution of microstructure, tensile properties, forming load, and die parameters during the process was investigated.

Keywords: constrained groove pressing, material properties, processing efficiency, strain homogeneity, die parameters

1. Introduction

Constrained groove pressing (CGP) is one of the most attractive SPD techniques available for fabricating ultra-fine grained (UFG) sheet or plane metallic materials. Since originally proposed by Shin et al. in 2002 [1], CGP has been successfully used for grain refinement and mechanical property improvement of various sheet metals and alloys. In this technique, the sheet sample is subjected to repetitive shear deformation via alternant pressings by asymmetrically groove dies and flat dies. Thus, a large strain can be uniformly accumulated throughout the whole sample without any significant change in its dimensions. As a result, a relatively homogeneous UFG structure can be obtained.

A schematic illustration of CGP process is shown in **Figure 1**. Before pressing, a set of asymmetrical groove dies (**Figure 1a**) and a set of flat dies (**Figure 1c**) are prepared. At first, the sample is pressed into the groove dies with a groove width, T, and a groove angle θ (**Figure 1a**). The gap

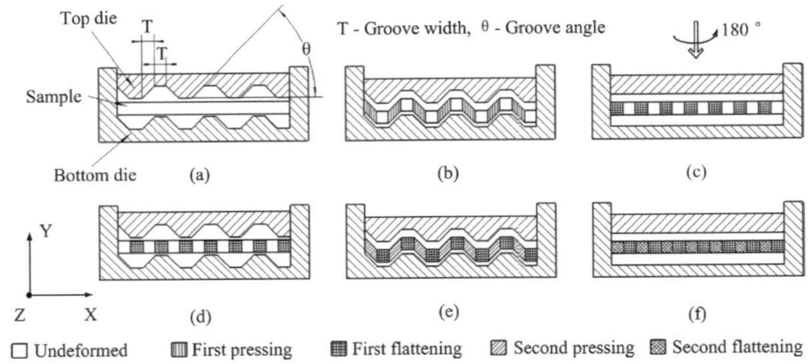

Figure 1. Schematic of constrained groove pressing (CGP).

between top and bottom dies is kept the same with the sample thickness. Thus, the inclined regions of the sample are subjected to pure shear deformation under plane strain condition while no deformation is induced in the flat regions (**Figure 1b**). Then, the grooved sample is placed between the flat dies and straightened (**Figure 1c**). Due to the tight constraint from the side walls of bottom die, the inclined regions previously deformed experience reverse shear deformation while the flat regions remain undeformed. After that, a rotation of the sample by 180° about Y-axis which is perpendicular to the sheet plane is performed (**Figure 1d**). This ensures the undeformed regions to be deformed during the second groove pressing (**Figure 1e**) and flattening (**Figure 1f**) due to the asymmetry of the groove dies. Thus, the alternate pressings with groove dies and flat dies result in a homogenous strain distribution throughout the sample without any changes in its dimensions. Generally, two groove pressings and two flattenings compose one CGP pass. Finally, a relatively uniform UFG structure can be obtained after a multi-pass CGP.

The theoretical equations for calculating the shear and effective strains accumulated for one pressing can be derived from **Figure 1**. A single pressing induces an engineering shear strain,

$$\gamma_{xy} = \tan\theta \tag{1}$$

where *H, T,* and θ are groove height, width, and angle, respectively. Shear strain is given as

$$\varepsilon_{xy} = \gamma_{xy}/2 \tag{2}$$

Since CGP is assumed as a pure shear deformation under plane strain condition [2], correspondingly, the effective strain,

$$\varepsilon_{eff} = \sqrt{\frac{4\varepsilon_{xy}^2}{3}} = \sqrt{\frac{4(\gamma_{xy}/2)^2}{3}} = \frac{\gamma_{xy}}{\sqrt{3}} = \frac{\tan\theta}{\sqrt{3}} \tag{3}$$

Therefore, the total effective strain accumulated in a CGP sample pressed by n passes is presented as

$$\varepsilon_{total} = n\frac{2\tan\theta}{\sqrt{3}} \tag{4}$$

Specifically, when the groove angle, θ is 45°, the groove width, T is equal to the groove height, H. Then, the engineering shear strain, $\gamma_{xy} = 1$, and the effective strain, $\varepsilon_{eff} = 1/\sqrt{3} \approx 0.58$. In this case, theoretically, one CGP pass induces a total effective strain of about 1.16 in the sample. Obviously, the groove angle, θ, directly determines the efficiency of strain accumulation.

CGP exhibits a great potential in producing UFG sheet metals, and until now, it has been successfully used for grain refinement and mechanical property improvement of various pure metals and alloys. From the process, significant influences on CGP by the structural parameters of groove dies including groove angle and width are expected. However, very limited studies are focused on this topic, and researchers still cannot reach an agreement on the influence rule. For instance, Borhani and Djavanroodi [3] investigated the effects of die design on a modified CGP process called rubber pad-CGP experimentally and numerically. They found that, compared with 45°, a higher groove angle of 50°could enhance the grain refinement and mechanical property improvement but reduce the strain homogeneity. Nevertheless, in another work on a covered sheet casing-CGP carried out by Sajadi et al. [4], it was demonstrated that an increase of groove angle from 45 to 53°could not promote more CGP passes and the improvement of mechanical properties. Peng et al. [5, 6]. analyzed the effects of groove width on CGP of a Cu-Zn alloy. They found that groove dies with the width increasing from 5 to 7 mm permitted more passes without crack formation but induced a lower rate of grain refinement and a slower increase in hardness.

As can be seen, the higher pass number resulted from a larger groove width is well explained by the above discussion, but the lower process efficiency still needs more detailed analysis. Thus, based on numerical and experimental methods, a multi-pass CGP was carried out on 1060 commercially pure aluminium, copper and Ni sheets in order to study the evolution of microstructure, tensile properties, forming load, die parameters and deformation mode during the process.

2. Finite element analysis and deformation homogeneity optimization of constrained groove pressing

2.1. FE-simulation and optimization

The FE-simulation model for CGP was built up using DEFORM-3D software. The initial dimensions of the sample are $100 \times 100 \times 2$ mm^3, and the material is aluminium alloy 5052. Both top die and bottom die are defined as a rigid body while the sample is plastic. During the deformation procedure, the bottom die is fixed and the top die moves in the $-Y$ direction at a constant speed. The coefficient of friction between the die surface and the sample is within a range (0.08–0.12) in cold forming of metals. The isothermal process is conducted at room temperature (20°C). The tetrahedral element is used for meshing and automatic re-meshing is activated. The total number of elements is 175,000 and the incremental step length is 0.2 mm. To ensure the accuracy, volume loss during simulation is taken into account. When the volume loss exceeds 5%, corresponding volume compensation is considered. The properties of the selected material are described by the following model equation of power law form:

$$\bar{\sigma} = c\bar{\varepsilon}^n \dot{\bar{\varepsilon}}^m + y \tag{5}$$

where $\bar{\sigma}$ is flow stress, $\bar{\varepsilon}$ is an effective plastic strain, $\dot{\bar{\varepsilon}}$ is effective strain rate, c is material constant, n is strain exponent, m is strain rate exponent, and y is initial yield stress value. For aluminium alloy 5052, fitted values of the parameters achieved from the software are $c = 56.4MPa$, $n = 0.0396$, $m = 0.0105$ and $y = 140MPa$.

In this study, one pressing cycle includes two groove pressings and two flattenings. The data of analysis points are extracted and analyzed after the sample is deformed after four cycles. In order to reduce the error, 5 mm from both edges of the sample along the longitudinal direction is ignored. The equivalent strain values of 300 points are measured uniformly as FE-simulation results along different paths within the range of 90 mm. The illustration of the analysis point selection is shown in **Figure 2**. The central line of the cross section at $Z = 0$ is identified as Path A, and the upper edge as Path B. The central line of the cross section at $Z = 50$ is defined as Path C. Among the paths concerned above, obviously, Path A is basically the most representative one for the deformation characteristics of the whole sample.

To describe the deformation homogeneity, I.F. is adopted as the index, which can be calculated by the following expression:

$$I.F. = \frac{\sqrt{\sum_{i=1}^{i=n} (H_i - \bar{H})^2/(n-1)}}{\bar{H}} \tag{6}$$

where n is the number of analysis points, H_i is the equivalent strain value of i-th point, $\bar{H}$ is the average strain value of all the analysis points. Normally, the lower I.F. value indicates more homogeneous deformation.

Four processing parameters were analyzed synthetically. They are groove width, groove angle, friction coefficient and deformation rate. The levels of each parameter are determined and

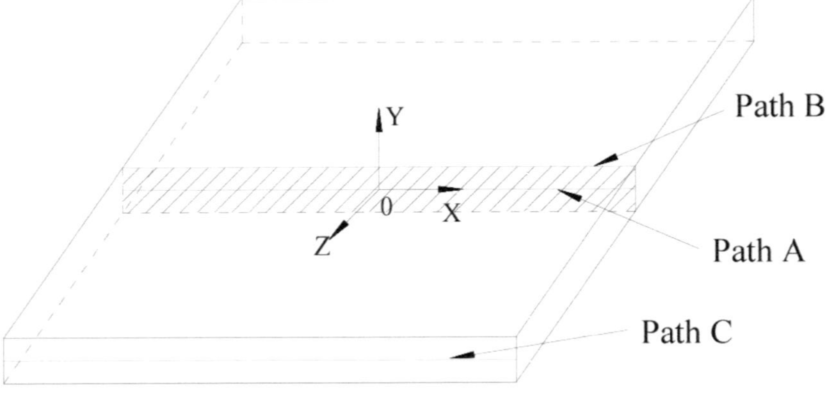

Figure 2. Extraction paths of analysis points.

shown in **Table 1**. The factors are expressed by A, B, C and D, and the levels by 1, 2 and 3. Due to the number of factors and levels chosen above, the L9 (3^4) orthogonal array is employed.

2.2. Results and discussion

The simulations of CGP for nine sets of parameter combination were conducted making use of DEFORM-3D, and values of equivalent strain for each plan were extracted along Paths A, B, and C of the deformed sample, respectively. After the I.F. value of each trial is figured out according to Eq. (6).

To examine the effect rule of the selected parameters, the mean S/N ratio for different levels of each factor are calculated, and the results are listed in **Table 2**. Since the smaller the better

Factors	Level 1	Level 2	Level 3
Groove width, A (mm)	1	1.5	2
Groove angle, B (°)	30	45	60
Friction coefficient, C	0.08	0.1	0.12
Deformation rate, D (mm/s)	0.16	1.6	16

Table 1. Analysis factors and levels.

Levels	Path A			
	A	B	C	D
1	26.293	24.317	28.805	28.433
2	31.176	30.882	30.094	31.005
3	31.592	33.861	30.162	29.623
Levels	**Path B**			
	A	B	C	D
1	26.417	24.114	27.441	28.606
2	30.742	29.838	30.120	30.200
3	30.247	33.453	29.845	28.599
Levels	**Path C**			
	A	B	C	D
1	26.172	18.841	25.315	25.289
2	26.832	28.505	28.598	26.685
3	26.834	32.491	25.925	27.864

Table 2. The mean *S/N* ratio for different levels of each factor.

quality characteristic was selected previously, higher value of S/N ratio implies a lower value of I.F. under the corresponding condition. Meanwhile, the CGP deformation effects are much better. Obviously, the I.F. value becomes lower as groove width, groove angle and friction coefficient increase except for deformation rate. Consequently, the best parameter combination is achieved and the combination is A3B3C3D2. When groove width is 2 mm, groove angle is 60°, the friction coefficient is 0.12 and deformation rate is 1.6 mm/s, the deformation is the most homogeneous.

The optimum result was compared with that of the initial model by applying the new parameter combination to the FE-simulation. During the initial process, the groove width is 2 mm, groove angle is 45°, the friction coefficient is 0.12 and deformation rate is 16 mm/s. It is evident that the mean amount of accumulative strain after optimization in the sample pressed by the same number of cycles is almost twice compared to the initial model. The effectiveness for grain refinement was significantly improved. Meanwhile, the difference between maximum and minimum values of equivalent strain almost keeps invariable before and after optimization. Regardless of the location of data points, the drop ranges of I.F. of equivalent strain are all about 50%, which means that the deformation homogeneity was enhanced distinctly after optimization. Furthermore, the I.F. value decreased from 0.028749 to 0.016103 and the S/N ratio rose from 30.828 to 35.866. As a result, the deformation turned to be more homogeneous.

3. Experimental investigation of pure aluminum sheets processed by constrained groove pressing

3.1. Experimental procedure

Cold-rolled 1060 commercially pure aluminum sheets with dimensions of 100 × 100 × 2 mm^3 were used. Before deformation, the sheets were annealed at 500°C for 4 h in an SX2-4-10 resistance-heated furnace utilizing high pure nitrogen as the protective atmosphere to ensure the annealing process.

In experiments, molybdenum disulfide (MoS_2) was coated on the surface of sheets as a lubricant. CGP process was conducted on a 5000 kN computer-controlled electro-hydraulic servo compression testing machine operated at a constant pressing speed of 5 mm·min-1 at room temperature. The pressing dies equipped with guide pillars and bushes had a groove angle of 45° and a groove width of 2 mm. Therefore, one CGP pass contained two groove pressings and two flattenings, yielding an effective strain of 1.16 throughout the whole sheet.

3.2. Results and discussion

A four-pass CGP with a total strain of 4.64 was successfully conducted on pure aluminum sheets. The microstructure evolution is shown in **Figure 3**. The grain size was measured based on OM observation and recorded in **Table 3**. The as-received material mainly consists of uniform and equiaxed grains with an average size of 29 μm (**Figure 3a**). Also, dislocation cells sized about 1 μm or more are evident in **Figure 3b**. Submicron and dislocation-free subgrains with well-defined boundaries begin to form at Pass 3 (**Figure 3c**). In **Figure 3d**, the subgrain

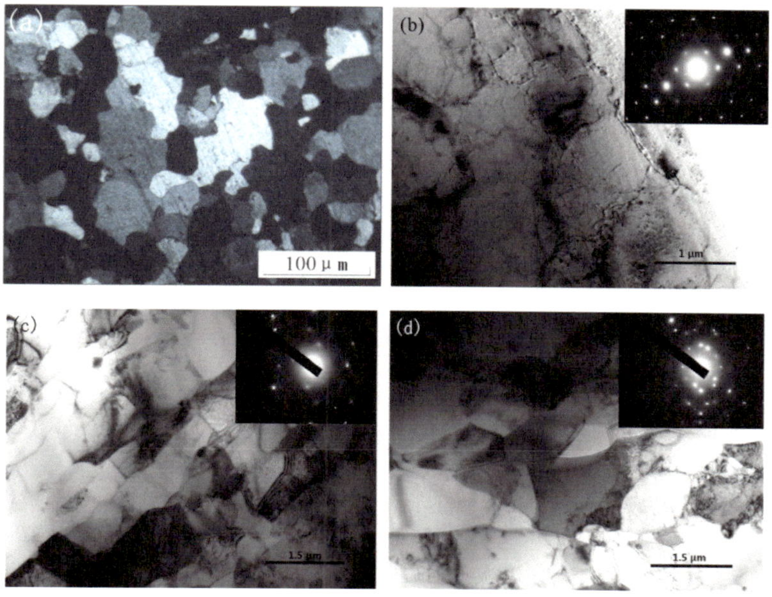

Figure 3. Optical micrograph for annealed sample (a) and TEM micrographs with their SAED patterns for (b) annealed, (c) three-pass and (d) four-pass samples, respectively.

Pass	0	1	2	3	4
Grain size (μm)	29.0	23.0	20.2	19.3	18.3
Yield strength (MPa)	23.8	93.8	98.2	103.6	93.5
Ultimate tensile strength (MPa)	66.9	101.1	109.1	111.5	96.4
Elongation to failure (%)	53.6	7.4	7.5	7.3	4.5

Table 3. Variations of grain size and tensile properties with pass number.

size undergoes a slight increase, and new tiny "grains" begin to appear along the boundaries. More diffused SAED pattern indicates higher misorientation angle between adjacent subgrains. Finally, the grain size estimated from OM observation is refined to 18 μm after Pass 4 and just 62% as that of the annealed material, as presented in **Table 3**.

Table 3 also lists the variations of tensile properties of aluminum samples with pass number. The ultimate tensile strength and yield strength increase rapidly to 101.1 and 93.8 MPa after Pass 1, respectively. However, in the following passes, they increase slowly. After that, the ultimate tensile strength and yield strength reach their maximum values, followed by reductions at Pass 4. This can be explained by the dynamic recovery of dislocations and subgrain coarsening shown in **Figure 3d**. Micro-cracks appeared on the sample surface during the later stages also contribute to the strength loss. The elongation decreases greatly from 53.6 to 7.4%. After that, it experiences a continuous and moderate reduction. No recovery of the elongation occurs due to flow softening, indicating that the effect of micro-cracks on tensile properties of material processed by CGP is more significant than flow softening [7].

For clarity, load-stroke curves of forming dies for Passes 1 and 2 are displayed in **Figure 4a** and **b**, respectively. Obviously, forming loads for all pressings can be divided into three distinct stages: rapid increase, moderate increase and second rapid increase. At the beginning of groove pressing, groove edges firstly contact and bend the sheet. After a short elastic deformation, the material begins to yield and comes to initial shearing around the groove edges, leading to the rapid increase of forming load. Then, as the upper die moves downwards, plastic deformation extends to other areas and results in the gradual increase of flow stress. Obviously, this stage covers more than one-half of the total pressing time and is the main stage of the three. A relatively steady load can be observed at this stage. At last, the grooves contact the entire surface of the sheet. The sheet is forced to be the same shape as the groove dies. Thus, another noticeable increase of forming load appears after the dies are fully closed. The load for flattening exhibits a similar variation tendency. Interestingly, a short plateau at the initial part of the second stage is observed from each flattening curve, as presented in the red circles in **Figure 4**. This is substantially attributed to the constraint by the container of flat dies. Before the extension of the plastic deformation to other areas, the sheet extends along the longitudinal direction firstly and is fully constrained by the side walls of the container.

In addition, as illustrated in **Figure 4a**, the stage division of load-stroke curves for the first groove pressing and flattening is not as clear as that for the other pressings because no strengthening was induced to the annealed aluminum sheets. It is concluded from the curves for all passes, not shown here, that the steady forming load in the second stage increases with CGP pass and a higher increasing rate is observed during the former passes. Within one pass, the steady loads for the two groove pressings are lower than those for the two flattenings which are almost equal to those for the groove pressings of the next pass. There are several contributing factors: (a) the strengthening of material increases with pass number and saturates at a high strain magnitude, (b) the inclined regions to be deformed in flattening have already been strengthened by shear deformation from the last groove pressing, and (c) the constraint by the side walls plays a more significant role in flattening than in groove pressing.

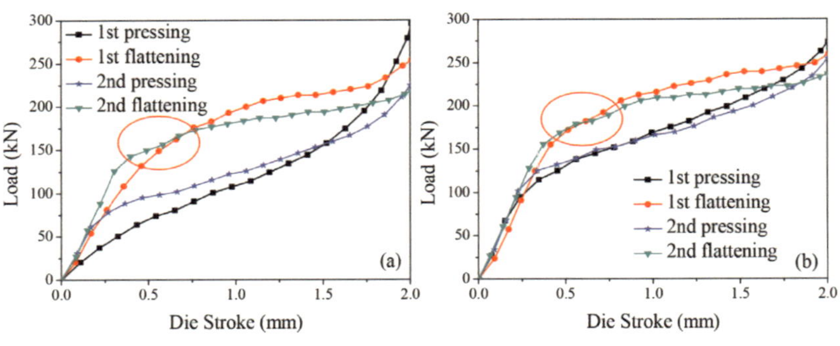

Figure 4. Load-stroke curves for (a) Pass 1 and (b) Pass 2, respectively.

4. Deformation efficiency and electrical resistivity of pure copper processed by constrained groove pressing

4.1. Experimental procedure

In the experiment, T2 commercially pure copper sheets were used. Before pressing, the sheets were annealed at 650°C for 2 h in an SX2-4-10 resistance-heated furnace with high pure nitrogen to avoid oxidation. The annealed sheets were machined to the dimensions of 100 mm × 100 mm × 2 mm. In this study, the samples underwent a four-pass CGP with a theoretically total strain of 4.64 until obvious cracks appeared on the surface of pressed samples at the last pass. The rest experimental conditions are the same with Section 3.1.

Resistances of sheets before and after CGP were measured using a four-electrode method by an H2ERM-1 resistance measuring device. A stable constant current of 500 mA provided by a PF66M digital multimeter passed through the specimen for approximately 4 s period to reduce Joule heating.

4.2. Processing efficiency

In this work, distinct cracks appear on the sheet surface at Pass 4, leading to uncompleted tensile tests. Thus, yield strength, ultimate tensile strength, and elongation to failure of materials for up to three passes are listed in **Table 4**. Great enhancement and rapid saturation of strengths are almost achieved after two passes, and further deformation induces significant decrease. Generally, this strengthening is introduced by work hardening and grain refinement. Previous papers proposed flow softening and microcracking as softening mechanisms [8–10].

Figure 5a and **b** illustrates the microstructures of pressed copper after one and three passes, respectively. In another experimental work on CGP of pure copper, only cell block structures with size of approximately 0.5 μm were obtained with a total strain of 3.48, even at cryogenic temperatures. In this study, as shown in **Figure 5a**, cell blocks with irregular shape exist as main structural features at the initial stages of straining. However, after three passes, subgrains with relatively distinct boundaries tend to be predominant in the structure, as indicated by the diffused SAED pattern in **Figure 5b**. The substructures are elongated and segmented into smaller ones, and the dislocation density inside is lower than before although the mean size remains the same. This seems to contribute to the flow softening mechanism concerned above. Meanwhile, **Table 4** also shows that the elongation decreases continuously as strain increases, and the highest rate of decrease is observed at Pass 1. No recovery of elongation happens during the last stages. Thus, it suggests that, during the whole CGP process, the drop of

Pass	0	1	2	3	4
Yield strength (MPa)	43.8	303.1	304.3	162.1	—
Ultimate tensile strength (MPa)	209.5	313.6	323.7	191.0	—
Elongation to failure (%)	51.5	6.1	5.7	3.0	—

Table 4. Tensile properties of pure copper samples.

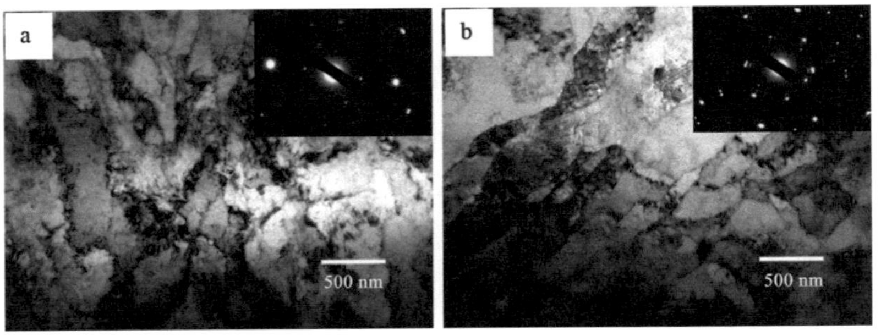

Figure 5. TEM micrographs and corresponding SAED patterns of copper after: (a) one and (b) three passes, respectively.

elongation caused by work hardening and microcracking always exceeds the rise due to grain refinement [11].

Interestingly, compared with previous research done by Krishnaiah et al. [8], the processing efficiency in our work is higher. For grain refinement, subgrains of about 0.5 μm with a lower dislocation density inside are obtained, instead of dislocation cells. OM observations show a mean grain size of around 21 μm with a more homogeneous distribution at the last pass. Correspondingly, tensile properties have experienced a large and rapid change from the initially low strengths and high elongation. Actually, the subgrains have already appeared at Pass 1 and the effective property improvement nearly saturates during this stage. Peng et al. have proposed a "bend-affected zone" in the pure shear region in their research [5]. The zone is induced by the extension of deformation around the groove corner. In this study, a smaller groove width of 2 mm and full constraint from the dies can enhance the effect of interface regions on the deformation of adjacent shear regions. This effect can be alleviated by the free elongation along the longitudinal direction of samples under the conditions of (unconstrained) groove pressing. Thus, we conclude that the processing rate is determined not only by the pass number but also by the die structure, such as groove width and constraint. In other words, die condition will influence the effective strain amount accumulated in the sample. For example, a small groove width and tight constraint indeed enhance the strain accumulation. However, due to more severe plastic deformation induced under this condition, initiation and propagation of micro-cracks occur much earlier, and consequently, the effective pass number is limited.

4.3. Electrical resistivity

In this research, the influence of SPD by CGP on the evolution of electrical resistivity of pure copper sheets has been investigated for the first time. The variation of electrical resistivity against pass number is presented in **Figure 6**. As can be seen, the annealed copper has an initial electrical resistivity of about 1.87 μΩ·cm, and it is positively correlated to the effective strain. The most rapid increase appears at the first pass. A decrease of the slope of variation curve is observed during the subsequent passes. The value of the electrical resistivity gets to the maximum of about 2.02 μΩ·cm after the third pass.

Crystalline defects such as impurities, dislocations and grain boundaries contribute to the resistivity of copper. SPD usually leads to the changes of grain size and dislocation density in

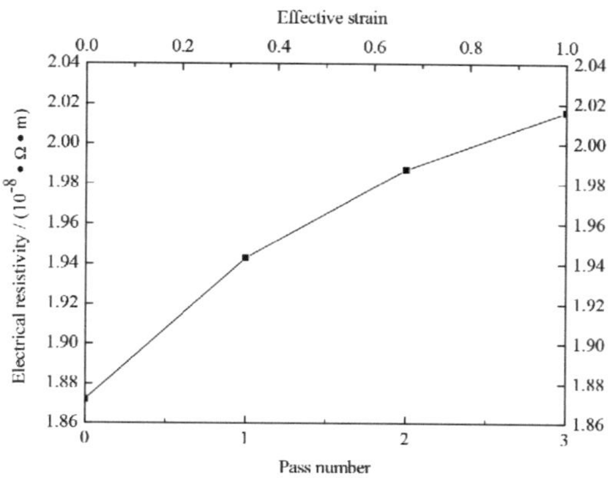

Figure 6. Variation of electrical resistivity against pass number.

metallic materials. As for pure metal, the microstructural effect on the electrical resistivity mainly comes from imperfections such as grain boundary and dislocation. Both of them play important roles in electron scattering, which will increase the electrical resistivity significantly [12].

During CGP, the microstructure of the material is characterized by noticeable grain refinement and high dislocation density. Refining the microstructure causes the increase of the total surface. Consequently, the electrical resistivity increases rapidly at the initial stages. Meanwhile, the occurrence of micro-cracks also contributes to the resistivity increase.

At a high level of plastic strain, there is a competition among dislocation recovery, homogeneity improvement and microcracking in deciding the electrical resistivity of materials. On one hand, recovery and rearrangement of dislocations happen at a larger magnitude of plastic strain, as indicated in **Figure 5b**, and slow down the dislocation proliferation and grain size reduction. Meanwhile, the improvement of microstructure distribution discussed above also helps to reduce the resistivity by homogenizing the various defects. On the other hand, micro-cracks nucleate and propagate on the sample surface, and the air gaps act as obstacles for electrons transport or scattering, resulting in the increase of electrical resistivity. Finally, a slightly lower increase rate of electrical resistivity is evident at Passes 2 and 3.

5. Influences of die structure on constrained groove pressing of commercially pure Ni sheets

5.1. Materials preparation and CGP experiments

Commercially pure Ni sheets with dimensions of $100 \times 66 \times 2$ mm^3 were used. Before pressing, the cold-rolled sheets were fully annealed at 750°C for 4 h in an SX2-4-10 resistance-heated furnace utilizing high pure nitrogen as the protective atmosphere to ensure the annealing process. The mean grain size estimated by line intercept method is approximately 28 μm.

Table 5 gives the schemes of experiments in this work. Qualitatively, considering the sample thickness of 2 mm, a lower groove width may induce severe shear in the sheets, and inefficiency is expected with a lower angle. Meanwhile, either a higher width or angle may change the deformation characteristics of CGP. In order to improve the research efficiency, representative schemes were conducted, as shown in **Table 5**. The groove dies used in the laboratory are shown in **Figure 7**. When pressing, the groove direction was perpendicular to the rolling direction (RD) of the sheets, and Teflon layers were used as a lubricant.

5.2. Influences of die structure on mechanical properties

In the experiments, CGP process was repeated until distinct cracks appeared on the surface of sheet samples. **Figure 8** gives the tensile properties of the sheets in different schemes. Obviously, the number of effective passes varies with die structure, and it is five, four and three in Schemes 1, 2 and 3, respectively. Thus, compared with previous works carried out by Satheesh Kumar and Raghu [13, 14], the pass number in this study is increased by the reduction of either groove angle or friction coefficient (by using Teflon layers as a lubricant). However, it should be noted that, according to Eq. (3), in Scheme 1 only a total effective strain of about 0.87 is imposed to the samples per pass.

In all the three schemes, CGP indeed greatly improves the strength of pure Ni sheets. The initially annealed pure Ni sheets have a yield strength of 79.6 MPa and ultimate tensile strength of 398.1 MPa. In Scheme 1, the yield strength rapidly increases to 407.9 MPa after Pass 1 and reaches the peak value of 439.7 MPa at Pass 3 with a strain of 2.61. Then, it experiences a slight fall to 406.9 MPa during the following two passes. In Scheme 2, the maximum yield

Die structure	$\theta = 30°$	$\theta = 37°$	$\theta = 45°$	$\theta = 53°$	$\theta = 60°$
$T = 1$ mm	–	–	S	–	–
$T = 2$ mm	S	S/E_1	S/E_2	S	S
$T = 3$ mm	–	–	S/E_3	–	–
$T = 4$ mm	–	–	S	–	–

Note: θ and T are groove angle and width, respectively. S represents numerical simulation; E_1, E_2 and E_3 represent Schemes 1, 2 and 3 in experiments, respectively.

Table 5. Schemes of experiments and numerical simulations.

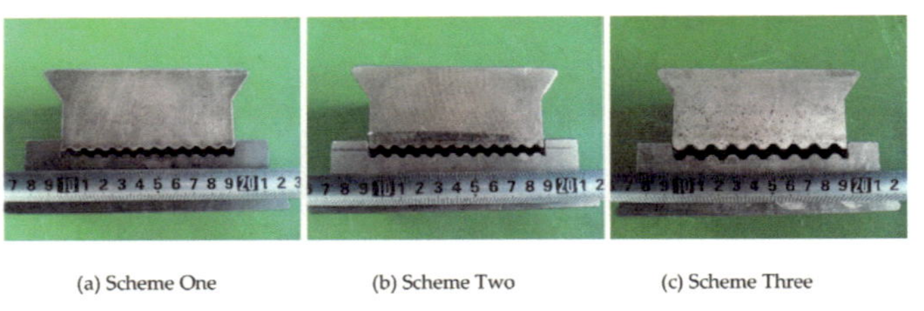

(a) Scheme One (b) Scheme Two (c) Scheme Three

Figure 7. Groove dies used in the experiments. (a) Scheme 1; (b) Scheme 2; (c) Scheme 3.

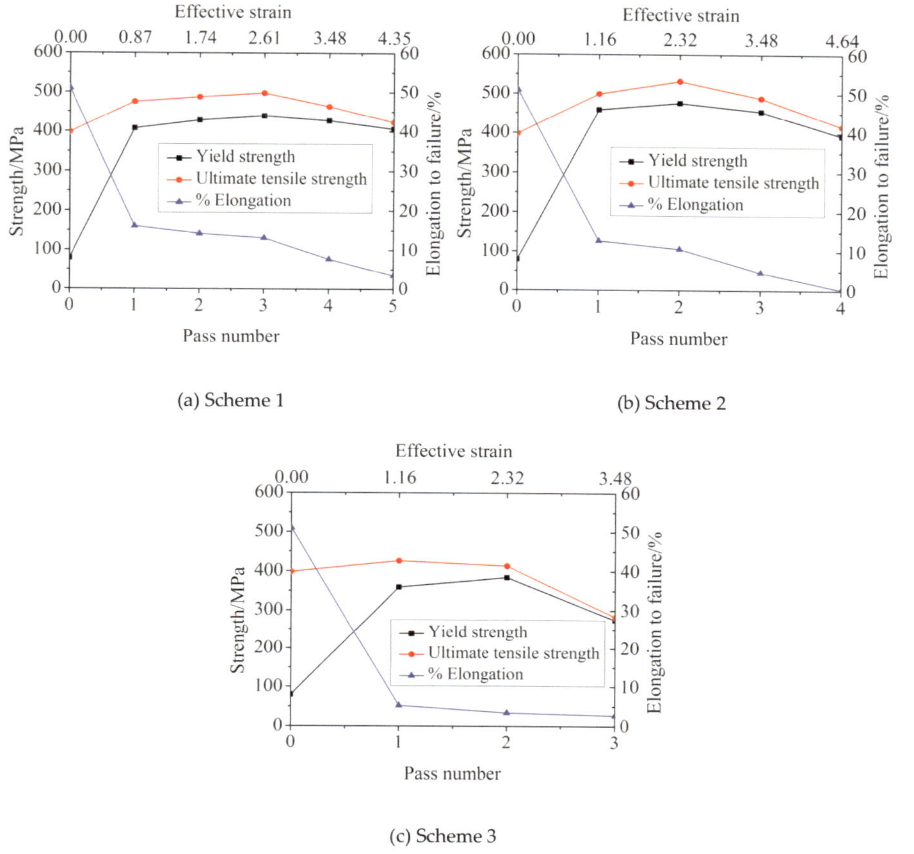

Figure 8. Tensile properties of Ni sheets before and after CGP. (a) Scheme 1; (b) Scheme 2; (c) Scheme 3.

strength of 476.3 MPa appears at Pass 2. After that, a decrease is observed, too. However, the highest strength is only 384.7 MPa at Pass 3 in Scheme 3. Meanwhile, it is observed that the evolution of tensile strength with pass number shows a similar trend to that of yield strength, but the increase of tensile strength at Pass 1 is not so significant, especially in Scheme 3. In addition, the elongation to failure of the annealed pure Ni sheets is about 51.0%, and all CGP samples experience a remarkable decrease at Pass 1. From Schemes 1–3, there is a gradual reduction in elongation of the processed materials, indicating a decrease of ductility caused by the increase of either groove width or angle [15].

The grain refinement and work hardening lead to the strength increase of materials during the initial stage of CGP, while the mechanisms of micro-cracking and flow softening contribute to the decrease at the later passes [10]. In this work, the optimum tensile properties of CGP pure Ni are obtained by the die design in Scheme 2. In Scheme 1, a lower groove angle relieves the intensity of shear deformation and also brings about more effective passes. Besides, the sheets processed in this scheme have a more acceptable ductility, but lower levels of total strain and strength than those in Scheme 2. Importantly, in Scheme 3, a larger groove width does not allow more CGP passes. Actually, materials get the worst results of property improvement in this scheme.

Figure 9 presents the evolution of average microhardness with pass number in different schemes. Generally, for all schemes, a sharp increase is observed from an initial value of 90.0 HV, and after that, only a slight increase is obtained. No reduction in hardness happens due to flow softening, indicating that micro-cracking plays a more important role than flow softening in the decrease of strength. Different from strength, the average hardness of CGP materials in Scheme 3 is higher than that in Scheme 1 during the three effective passes. However, at Pass 5 in Scheme 1, the final hardness catches up with that at Pass 3 in Scheme 3. This suggests the importance of strain amount in the enhancement of mechanical properties. In addition, despite the same rate of strain accumulation in Schemes 2 and 3, a higher level of hardness is observed in Scheme 2. Thus, it is believed that, besides strain amount, there must be other factors based on the change of die structure accounting for this result. And this will be discussed in the following sections.

Briefly in this work, with a groove width of 2 mm and a groove angle of 45° in Scheme 2, pure Ni sheets with a thickness of 2 mm processed by a two-pass CGP acquire the best mechanical properties. The yield and tensile strengths are 476.3 and 532.3 MPa, respectively, the elongation to failure is 10.6%, and the average microhardness is 218.9 HV.

5.3. Influences of die structure on microstructure

Figure 10 shows the TEM micrographs and corresponding SAED patterns of CGP pure Ni sheets at the last pass in different experimental schemes. In **Figure 10a**, only irregular dislocation cells can be found in the microstructure even processed with a total strain of 4.35. The cell structures with high dislocation density both inside and at the boundaries are not homogeneous. Similar substructures appear in Scheme 3, as illustrated in **Figure 10c**. However, the less diffused SAED pattern indicates lower misorientation angles between cells in Scheme 3 due to a smaller strain amount of 3.48.

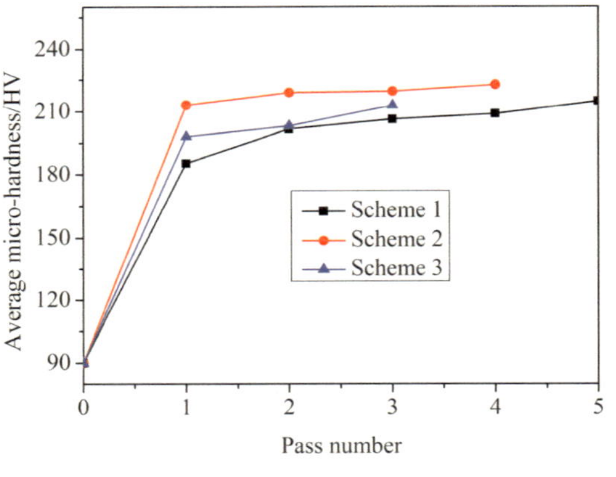

Figure 9. Average micro-hardness of Ni sheets before and after CGP.

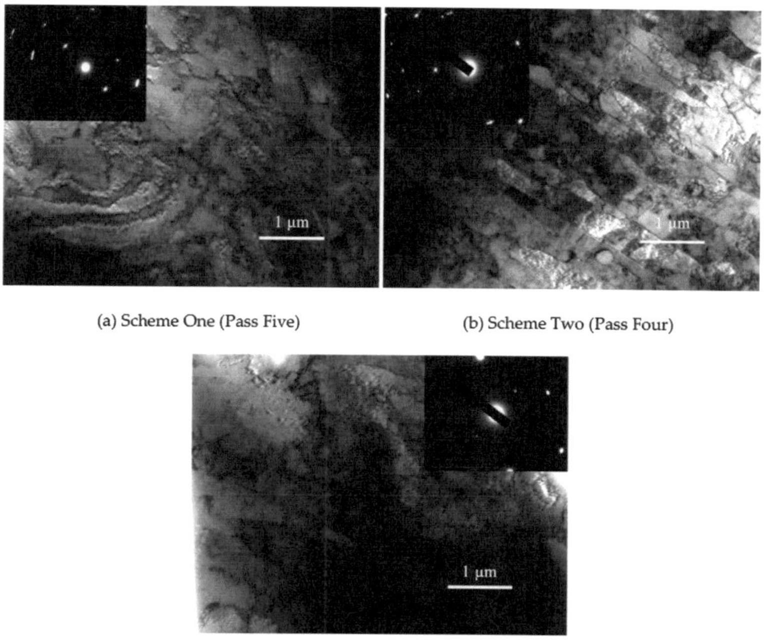

(a) Scheme One (Pass Five) (b) Scheme Two (Pass Four)

(c) Scheme Three (Pass Three)

Figure 10. TEM micrographs and corresponding SAED patterns of CGP Ni sheets. (a) Scheme 1 (Pass 5); (b) Scheme 2 (Pass 4); and (c) Scheme 3 (Pass 3).

Corresponding to the optimum mechanical properties, the die design in Scheme 2 leads to the best results of grain refinement. As shown in **Figure 10b**, a large number of elongated subgrains come into being with a mean width of about 500 nm. Separate grains with clear boundaries and high misorientation angles indicated by the SAED pattern are obtained. Compared with pure Cu [11], the material in this work experiences a higher efficiency of grain refinement due to its intermediate SFE.

5.4. Influences of die structure on texture evolution

Small crystalline size and lattice distortion are two principal imperfections in crystalline materials and usually cause peak broadening in XRD patterns compared with those attained from perfect crystal diffraction. Thus, the evolutions of cell size and lattice micro-strain can be reflected by the deviation of line profile from perfect diffraction. In this work, XRD patterns of pure Ni sheets before and after one CGP pass in different experimental schemes were examined and presented in **Figure 11**. Clearly, diffractions from four crystalline planes are more intense than those from others, and they are (111), (200), (220) and (311). Considering the diffractions of these peaks on all samples, the microstructure evolution can be analyzed. Compared with the annealed pure Ni sheets, the peaks exhibit different degrees of broadening after Pass 1, suggesting various results of grain refinement. Specifically, the most and least distinct broadenings can be carefully observed in Schemes 2 and 3, respectively. This confirms the TEM observations of microstructure evolution discussed above.

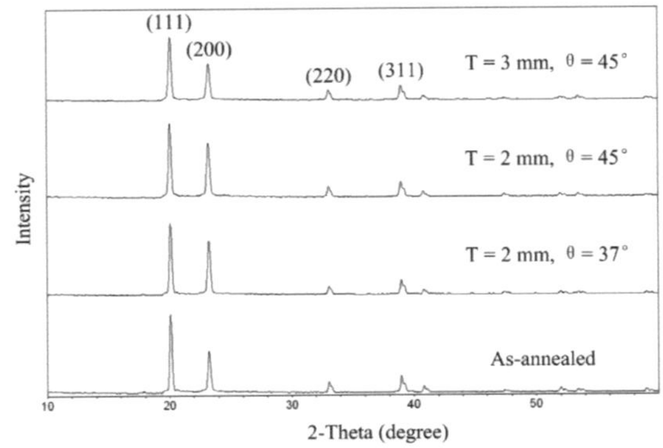

Figure 11. XRD patterns of Ni sheets before and after Pass 1.

6. Conclusions

1. The best parameter combination is A3B3C3D2 under the established conditions. The deformation is the most homogeneous when groove width is 2 mm, groove angle is 60°, the friction coefficient is 0.12 and deformation rate is 1.6 mm/s. The I.F. value of optimum model drops by about 50% and the deformation homogeneity is greatly improved. Compared with the initial model, plastic strain accumulated in samples deformed by the same number of passes increases to almost twice. The deformation efficiency of CGP technique is promoted significantly.

2. A four-pass CGP was carried out on commercially pure aluminum sheets. The grain size is refined from 29 μm of the annealed sample to 18 μm after four passes. Dislocation-free subgrains of the submicron level with well-defined boundaries are obtained. The ultimate tensile strength and yield strength have been improved significantly. All the load-stroke curves can be divided into three stages: rapid increase, moderate increase and second rapid increase. During flattening, a short plateau appears at the beginning of the second stage.

3. For commercially pure copper sheets, the grain size is reduced from 30 μm of the annealed material to around 21 μm after CGP deformation, and subgrains sized about 0.5 μm with distinct boundaries are obtained at Pass 3. The sharp changes of microstructure and mechanical properties indicate a high processing efficiency of this technique with a small groove width of 2 mm and tight constraint. The electrical resistivity of pure copper exhibits a near-linear increase with the straining. Crystalline imperfections, microcracking and microstructure uniformity together determine the evolution of electrical resistivity.

4. For pure Ni sheets, the optimum mechanical properties of pure Ni sheets with a thickness of 2 mm are obtained by a groove width of 2 mm and a groove angle of 45° after two CGP passes. In the experiments, a lower groove angle eases the intensity of shear deformation and permits more effective passes while the process efficiency is reduced. Besides, a higher groove width cannot induce more passes but obtain the worst CGP results. In both

cases, only cell structures with high dislocation density are observed. The peak broadenings in XRD patterns of pure Ni sheets after CGP confirm the results of grain refinement indicated by TEM observations: Scheme 2 > Scheme 1 > Scheme 3.

Acknowledgements

The research work was supported by the National Natural Science Foundation of China (51375269, 51675307).

Author details

Yanjin Guan[1]* and Zongshen Wang[2]

*Address all correspondence to: guan_yanjin@sdu.edu.cn

1 Key Laboratory for Liquid-Solid Structural Evolution and Processing of Materials (Ministry of Education), Shandong University, Jinan, China

2 School of Mechanical Engineering, Shandong University of Technology, Zibo, China

References

[1] Shin DH, Park J, Kim Y, Park K. Constrained groove pressing and its application to grain refinement of aluminum. Materials Science and Engineering: A. 2002;**328**:98-103

[2] Shirdel A, Khajeh A, Moshksar MM. Experimental and finite element investigation of semi-constrained groove pressing process. Materials & Design. 2010;**31**:946-950

[3] Borhani M, Djavanroodi F. Rubber pad-constrained groove pressing process: Experimental and finite element investigation. Materials Science and Engineering: A. 2012;**546**:1-7

[4] Sajadi A, Ebrahimi M, Djavanroodi F. Experimental and numerical investigation of Al properties fabricated by CGP process. Materials Science and Engineering: A. 2012;**552**:97-103

[5] Peng K, Zhang Y, Shaw LL., Qian KW. Microstructure dependence of a Cu-38Zn alloy on processing conditions of constrained groove pressing. Acta Materialia. 2009;**57**:5543-5553

[6] Peng K, Su L, Shaw LL, Qian KW. Grain refinement and crack prevention in constrained groove pressing of two-phase Cu-Zn alloys. Scripta Materialia. 2007;**56**:987-990

[7] Wang ZS, Liang P, Guan YJ, Liu YX, Jiang LB. Experimental investigation of pure aluminum sheets processed by constrained groove pressing. Indian Journal of Engineering and Materials Sciences. 2014;**21**:121-127

[8] Krishnaiah A, Chakkingal U, Venugopal P. Applicability of the groove pressing technique for grain refinement in commercial purity copper. Materials Science and Engineering: A. 2005;**s410–s411**:337-340

[9] Khodabakhshi F, Kazeminezhad M, Kokabi AH. Constrained groove pressing of low carbon steel: Nano-structure and mechanical properties. Materials Science and Engineering: A. 2010;**527**:4043-4049

[10] Hosseini E, Kazeminezhad M. Nanostructure and mechanical properties of 0-7 strained aluminum by CGP: XRD, TEM and tensile test. Materials Science and Engineering: A. 2009;**526**:219-224

[11] Wang Z.S, Guan YJ, Liang P. Deformation efficiency, homogeneity, and electrical resistivity of pure copper processed by constrained groove pressing. Rare Metals. 2014;**33**:287-292

[12] Khodabakhshi F, Kazeminezhad M. The effect of constrained groove pressing on grain size, dislocation density and electrical resistivity of low carbon steel. Materials & Design. 2011;**32**:3280-3286

[13] Satheesh Kumar SS, Raghu T. Tensile behaviour and strain hardening characteristics of constrained groove pressed nickel sheets. Materials & Design. 2011;**32**:4650-4657

[14] Satheesh Kumar SS, Raghu T. Mechanical behaviour and microstructural evolution of constrained groove pressed nickel sheets. Journal of Materials Processing Technology. 2013;**213**:214-220

[15] Wang ZS, Guan YJ, Wang GC, Zhong CK. Influences of die structure on constrained groove pressing of commercially pure Ni sheets. Journal of Materials Processing Technology. 2015;**215**:205-218

Chapter 4

High-Pressure Torsion: Experiments and Modeling

Marina Borodachenkova, Wei Wen and
António Manuel de Bastos Pereira

Additional information is available at the end of the chapter

http://dx.doi.org/10.5772/intechopen.69173

Abstract

The high-pressure torsion (HPT) process has been the subject of many investigations as a new method of processing for nanostructured materials due to its ability to develop nanostructures with high-angle grain boundaries. This chapter examines the various publications describing the experimental studies of the effect of HPT on the mechanical behaviors and alterations of microstructural features in applications to various pure and alloyed metals. Moreover, an overview of the modeling approaches developed through the last decade, considering the main advantages/limitations, is analyzed.

Keywords: high-pressure torsion, grain refinement, modeling

1. Introduction

The grain size is one of the essential factors controlling the mechanical and physical properties of polycrystals. It is well known that the strength of polycrystalline materials can be improved by reducing the grain size. Materials with fine microstructure usually possess extraordinary properties, including high strength, good toughness, and long fatigue life [1]. For this reason, producing metals with a very small grain size has attracted wide interest recently. To achieve materials with ultrafine-grained (UFG) structures and superior mechanical properties, severe plastic deformation (SPD) has emerged as the fundamental process, as pointed out in review articles by Mazilkin et al. [2] and Zhu et al. [3].

Synthesis of ultrafine-grained (UFG) materials by severe plastic deformation (SPD) refers to various experimental metal forming procedures that may be applied to impose very high strains on materials leading to exceptional grain refinement. One of the most important features of SPD processing is that the shape of the sample is retained by using special tool geometries, which:

1. Prevents the free flow of the material and thereby produces a significant hydrostatic pressure.

2. Allows imposing strain without any significant change in the overall dimensions of the sample. Therefore, it is possible to repeat SPD process on the sample to achieve extreme large strain.

In the early studies of UFG microstructure formation during SPD, two processing methods have been investigated more intensively: equal channel angular pressing (ECAP) and high-pressure torsion (HPT). These methods have been applied to a wide range of materials: pure metals, alloys, composites, and ceramics. Compared to ECAP, HPT is especially effective to introduce extremely large shear strain which triggers strong grain refinement. In the present chapter, the evolution of the microstructural and mechanical properties during HPT is summarized. Following that, the overview of the modeling approaches applied to predict the microstructure evolution and the stress-strain distribution during HPT is presented.

2. The HPT procedure

A brief introduction of the HPT procedure is presented in this section. More detailed descriptions are available in the literature [2, 3]. The principle of the modern HPT process is illustrated schematically in **Figure 1** [4]. A specimen is held between the plunger and the support and is strained in torsion under the applied pressure (*P*) in the order of several GPa (1–10 GPa). A lower holder rotates and deforms the specimen by the contact surface friction forces so that deformation proceeds under a quasi-hydrostatic pressure. In practice, there are two main types of HPT processing depending on the shape of the anvils: the unconstrained (**Figure 2a**) and the constrained (**Figure 2b** and **c**) HPT. In unconstrained HPT, samples are placed between two anvils and subjected to HPT processing. In such a case, the sample material is free to flow outward when the

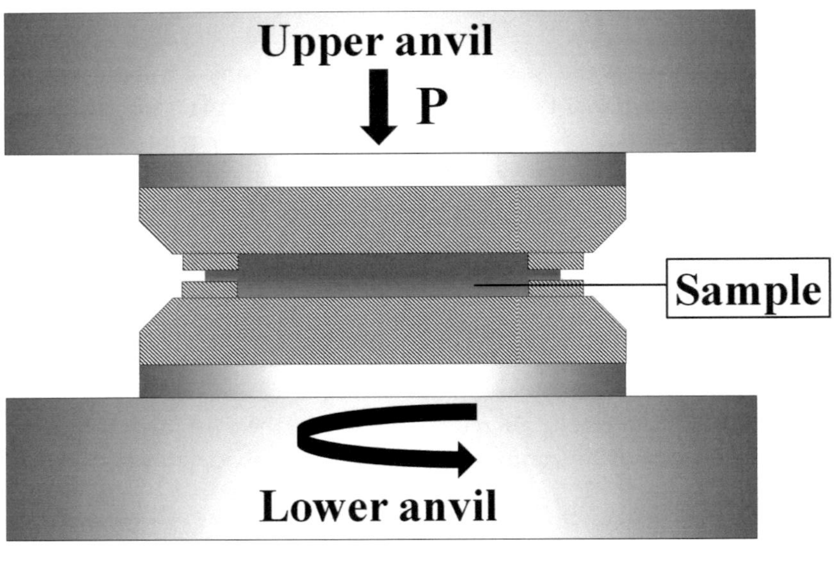

Figure 1. A schematic view of the HPT setup.

high pressure is applied. Samples are thus much thinner after HPT processing [6]. In constrained HPT, samples are placed into the cavity of the lower anvil or both anvils [6], which can prevent material flowing outward. Therefore, the thickness reduction is not evident during HPT. Normally, the constrained HPT is a more common method since this designing is conducted with a more effective back-pressure to the samples [7–9]. However, it is generally difficult to achieve an idealized constrained condition. The experiments are often performed under a quasi-constrained condition where there is at least some limited outward flow between the anvils.

For an infinitely small rotation, $d\theta$, and a displacement, dl, it follows from **Figure 3** that where r is the radius of the disk, the incremental shear strain, $d\gamma$, is given by:

$$d\gamma = \frac{dl}{h} = \frac{rd\theta}{h} \tag{1}$$

where h is the disk thickness.

By further assuming that the thickness of the disk is independent of the rotation angle θ, it follows from formal integration that since $\theta = 2\pi N$, the total shear strain, γ, can be expressed as:

$$\gamma = \frac{2\pi N r}{h} \tag{2}$$

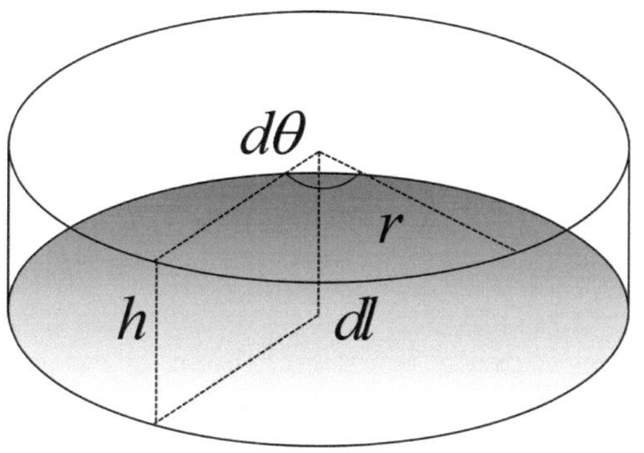

Figure 2. A schematic view of the sample dimension and the parameters used to estimate the imposed strain in HPT.

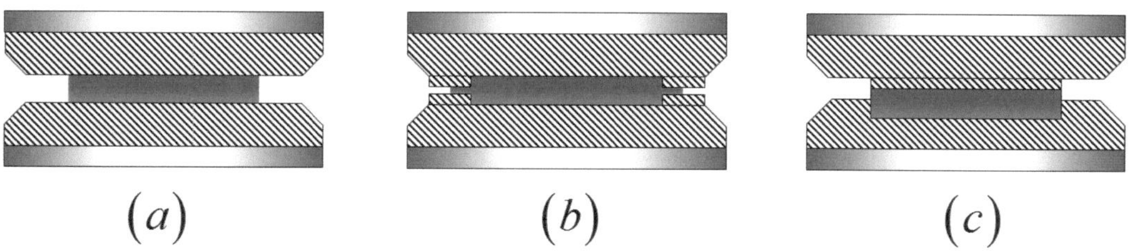

Figure 3. A schematic view of the (a) unconstrained and (b and c) constrained HPT processing conditions.

where N is the number of rotations. The equivalent Von Mises strain is calculated using a commonly used relationship:

$$\varepsilon = \frac{\gamma}{\sqrt{3}} \tag{3}$$

Theoretically, the imposed strain during HPT is given by Eq. (2). As a result of this expression, the strain is equal to zero at the center of the sample and increases linearly until reaching a maximum near the edges. Thus, the microstructure produced by HPT is heterogeneous. Some authors reported that the microhardness also varies significantly along the radius of disks processed by HPT [10–12].

Nevertheless, it is possible to obtain homogeneous structures along disk diameter by increasing the number of revolutions [4, 13]. For example, in the work of Xu et al. [14], high purity aluminum disks were processed by HPT at room temperature under pressures of 1.25, 2.5, and 6 GPa for 1, 3, and 5 turns. It has been reported that at the early stages of deformation, the hardness at the disk center is higher than that at the edges, and the hardness becomes homogeneous with high level of deformation. This kind of material response has been observed for a material where recovery is rapid, as in the pure aluminum used in this investigation. By contrast, for a material where recovery is slow, the hardness is initially lower in the center, but gradually the microstructure evolves into a homogeneous condition.

In the work of Kawasaki et al. [15], processing by high-pressure torsion has been conducted through 1/4, 1, and 5 turns, and detailed microhardness measurements were recorded on high purity (99.99%) aluminum. The hardness is initially high in the centers of the HPT disks but it decreases with torsional straining to become reasonably homogeneous.

3. Influence of HPT on the mechanical and microstructural properties

The HPT process has been the subject of many investigations due to its ability to develop homogeneous nanostructures with high-angle grain boundaries [16]. The mechanical behavior and microstructural feature evolutions have been extensively studied for a wide range of pure metals and alloys [4, 5, 13, 17–28]. In this section, the overview of the nanosized microstructure formation and the improvements in the mechanical properties during HPT are discussed.

3.1. Grain refinement

A tremendous amount of experimental works has been published on grain refinement by HPT. It has been reported that there is an ultimate minimum grain size that can be achieved by HPT. The results of the grain refinement for various metals and alloys during HPT are summarized in **Table 1**. As it can be seen, refinement ratio varies from 100 to 1800.

Despite the intensive empirical studies on the formation of the ultrafine-grained microstructures, only several studies have attempted to understand the mechanisms of the grain

http://dx.doi.org/10.5772/intechopen.69173

Metal/alloys	Initial grain size (μm)	Saturated grain size (nm)	Yield stress (MPa)	Pressure (GPa)	Number of rotations	References
Commercial 7075 Al alloy		26	1000	6	10	[29]
Commercial pure Ti	2–10	30	–	5	5	[30]
High purity (99.99%) Ni	100	170	Hv = 3.2 GPa	6	5	[4]
Al-7075 alloy	Extruded lengths up to ~450 μm and widths of ~8 μm	500 (disk center) 250 (disk edge)	Hv~230	6	10	[31]
Cu-0.1 wt% Zr	20	270 (disk center) 230 (disk edge)	–	6	10	[32]
Al-1% Mg	400	230	Hv~110	6	10	[33]
High purity copper	40	200–250	Hv~140	6	10	[34]
Al-5 wt% Fe alloy (metastable supersaturated solid solution)		150		5		[35]
Armco-iron 99.95%	40	100	Hv=4.6 GPa	7	5	[36]
Mg-3%, Al-1% Zn	150–200	150–200	Hv~110	2.5	15	[37]
Pure Al (99.7%)	200–500	800	Hv~600MPa	1	8	[17]
Pure tantalum (99.9%)	60	160	Hv~400 σy = 1300MPa	6	10	[38]
Pure Ti (99.99%)	10	150	Hv~248	2	5	[39]
Pure Zr (99.9%)	30	100	Hv~380 σy = 1150MPa	6	20	[22]
Pure hafnium (99.99%)		180	Hv~360	4	10	[40]

Table 1. The grain refinement during HPT for various metals/alloys.

refinement. Most of the theories are based on the idea that a high dislocation density is introduced in the material due to heavy straining and later the dislocations rearrange into ultrafine grain structures, as illustrated in **Figure 4**. In addition to the dislocation mechanisms for ultrafine-grained microstructure formation, some other mechanisms of the grain refinement have been described in the literature. For example, Isik et al. [41] reported that, in the case of Co-Cr-Mo alloy, the deformation-induced phase transformation ($\gamma \rightarrow \varepsilon$) contributes to grain refinement via the formation of ε platelets which subdivide the γ grains. Borodachenkova et al. [42] demonstrated that in Al-Zn alloy, the intense dislocation pinning between Zn precipitates within Al grains produces an effective division of the grains and accelerates the grain refinement process. The work of Liu et al. [43] indicated that the microtwins can further promote division and break down the grains into subgrains.

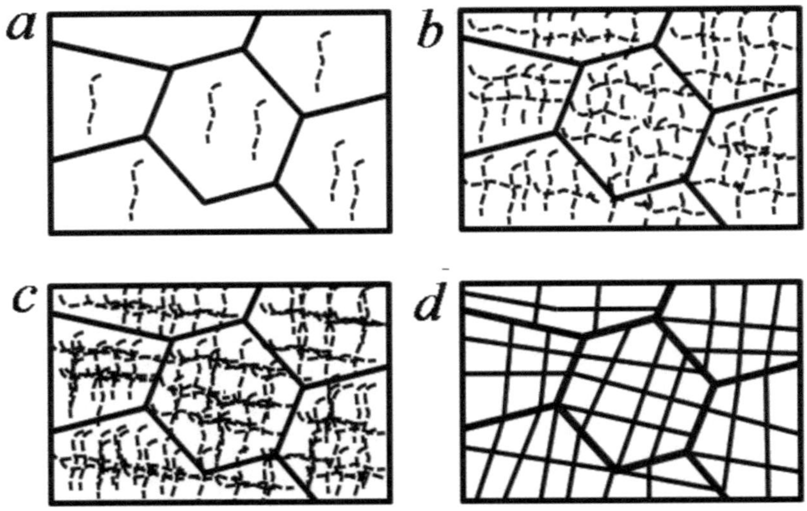

Figure 4. A schematic view of grain refinement during HPT, describing sequentially the processes of (a) the generation/accumulation of dislocations, (b) the formation of subgrain boundaries, (c) the increase in the misorientation angle, and (d) the division of grains into subgrains.

3.2. Improvements in mechanical properties

As it has been widely reported, HPT processing leads to strong grain refinement, and according to the well-known Hall-Petch relation, the material strength is inversely proportional to the grain size. However, the recent works reported that HPT might lead to strain softening and grain refinement simultaneously for pure metals as well as for the alloys. For example, in the work of Ito et al. [44], the purity of Al can evidently influence the material hardness. With increasing Al purity, the grain size dependence of hardness becomes less significant. For ultrahigh pure 6NAl, the hardness variations with respect to the grain size follow an inverse Hall-Petch relationship. The hardness in the HPT-processed state becomes lower, although the grain size (~20 μm) is smaller than the non-deformed state (larger than 1 mm). The main reason for the observed behavior is that the high-angle grain boundaries act mainly as dislocation sinks (the dislocations moved fast and disappeared in high-angle grain boundaries) in ultrahigh pure Al.

Earlier, Ito and Horita [24] investigated the evolution of the mechanical behavior of pure aluminum during the HPT process. The initial grain size before HPT was 250 μm. After one rotation under hydrostatic pressure of 6 GPa, the average grain size was reduced to approximately several microns. The results show that the hardness of pure Al initially increases with increasing strain and then decreases to a saturation value. Based on the transmission electron microscopy (TEM) observations, the following explanation for the softening behavior has been suggested: in the region where the hardness increases, the dislocation accumulation and the subgrain boundary formation occur. The increase in hardness is attributed to

	High-pressure torsion				Superplasticity			References
Alloy	D (μm)	d_{sat} (nm)	N	P (GPa)	T_t (K)	$\dot{\varepsilon}$ (s^{-1})	L_{max} (%)	
Ti-6Al-7Nb	5	100	5	6	1073K (~0.58T_m)	2×10^{-3}	930%	[48]
Mg-8%wt Li		500	5	3	298	1×10^{-3}	310	[49]
					323		400	
					343		780	
					423		1010	
					473		1330	
Al-2024	~33	~240	5	6	573	2×10^{-2}	520	[50]
						1×10^{-2}	470	
						5×10^{-3}	580	
						2×10^{-3}	460	
						1×10^{-3}	350	
Al-2024	~33	~240	5	6	623	2×10^{-2}	410	[50]
						1×10^{-2}	710	
						5×10^{-3}	470	
						2×10^{-3}	640	
						1×10^{-3}	650	
Al-2024	~33	~240	5	6	723	2×10^{-2}	220	[50]
						1×10^{-2}	370	
						5×10^{-3}	170	
						2×10^{-3}	380	
						1×10^{-3}	430	
Mg-9% Al-1% Zn	30	1500	10	3	573	1×10^{-1}	410	[51]
						1×10^{-2}	860	
						1×10^{-3}	1050	
						1×10^{-4}	1308	
Al-7075+10% vol Al_2O_3 metal matrix composite	8	300	20	6	623	1×10^{-1}	305	[52]
						1×10^{-2}	595	
						1×10^{-3}	345	
Ti-6Al-4V	9.5	77±15	20	6	673	5×10^{-3}	540	[53]
						1×10^{-3}	540	
						5×10^{-4}	440	
						1×10^{-4}	790	

Table 2. The superplasticity properties after HPT processing.

an increase in dislocation density which causes more chances of the mutual interaction of dislocations within grains and of blocking of dislocation motion by the presence of subgrain boundaries. Later, the dislocation density in the subgrains starts to decrease due to the dislocation annihilation at subgrain boundaries. Meanwhile, this annihilation leads to an increase in the misorientation angles, which will further promote the dislocation absorption at the boundaries [24]. This is the reason why the misorientation increases with straining, and more grains are surrounded by higher-angle boundaries. Finally, hardness saturates at a constant level when the dislocation accumulation is balanced with the dislocation absorption at high-angle boundaries. Pang et al. [45] observed that the HPT processing leads to strain softening for the Cu-Al alloys with SFE higher than 28 mJ/m^2, in which dynamic recovery is more noticeable during plastic deformation. For alloys with the lower SFE of 6 mJ/m^2, the strain softening has been restrained, and strain hardening played a dominant role in the deformation process. Mazilkin et al. [2, 46] demonstrated that the HPT of the Al-Zn (10/20/30%wt) and Al-Mg (5/10%wt) alloys leads to a strong grain refinement (from 15 μm to 370 nm) and decomposition of supersaturated solid solution of Zn and Mg in Al. The decomposition of supersaturated solid solution results in material softening.

3.3. Low-temperature superplasticity

Superplasticity refers to the capability of a polycrystalline metal to perform an elongation of at least 400% in tension load and with an associated value for the strain rate sensitivity. It is well established that the superplastic properties are related to grain refinement of materials [47]. Necessary conditions for the superplasticity are (1) stable fine-grained microstructure (grain size less than 10 μm) and (2) temperature higher than half of the melting temperature. HPT processing is an effective method for grain refinement to the submicrometer or even nanometer level. It is also widely reported that the enhanced superplastic properties have been obtained by HPT processing at relatively low temperatures. The recent results for the superplasticity properties after HPT are summarized in **Table 2** (D—initial grain size, d_{sat}—saturated grain size, N-—number of rotations, P—applied pressure during HPT processing, T_t—testing temperature, T_m—melting point, $\dot{\varepsilon}$—strain rate, and L_{max}—maximum elongation).

4. Modeling

Despite the large quantity of studies performed on HPT, most of them are only dedicated to microstructural and mechanical characterization. Recently, some researchers attempted to develop dislocation-based models to capture microstructural evolutions and the resulting changes in properties under large strain. Some other models, which are rooted in the finite-element method, mainly focus on the description of the stress/strain distribution and evolution along the sample radius, using a more empirical approach to describe the mechanical response without getting into the details of the micromechanisms. This section aims to provide an overview of the modeling approaches developed in the last decade, with a discussion of their main advantages/limitations.

4.1. Microstructure-based approaches

The microstructure-based models, describing the grain refinement due to large strain (particularly under HPT), are usually based on the notion that the dislocation cell walls, which form in the early step of the deformation, transform gradually to high-angle grain boundaries. This type of models are commonly based on the approach of Kocks and Mecking [54], which describes the deformation behaviors of pure metals and alloys via a single internal variable, namely, the total dislocation density ρ_{total}. Estrin [55] proposed a constitutive model to express the hardening behaviors of cell-forming crystalline materials at large strains. A dislocation structure that is developed under torsion deformation can be considered as cellular, with subgrain boundaries containing a high dislocation density separating subgrain interiors where the dislocation density is significantly lower. The volume fraction of the walls f_w is calculated using the following expression:

$$f_w = \frac{2\omega d - \omega^2}{d^2} \tag{4}$$

where ω is the wall thickness and d is the subgrain size, which is proportional to the average dislocation interspacing, $(\propto 1/\sqrt{\rho_{total}})$. The total dislocation density is determined as:

$$\rho_{total} = f_w \rho_w + (1 - f_w)\rho_c \tag{5}$$

where ρ_c is the dislocation density in the cell interior dislocation density, whereas ρ_w is the dislocation density in the cell walls.

The macroscopic stress τ is considered as the sum of the stresses within the walls and cell interiors:

$$\tau = f_w \tau_w + (1 - f_w)\tau_c \tag{6}$$

To validate this model, it has been applied to predict the torsion deformation of pure copper. The predicted hardening curve is compared with experimental results on copper torsion, and a good agreement between theory and experiment is achieved. Since other mechanisms are not accounted for, this model is restricted to the material in which the dislocation hardening is the dominant mechanism.

Zhang et al. [56] has developed a microstructural model that is based on the evolution of geometrically necessary dislocations (GND) and statistically stored dislocations (SSD) that incorporate grain refinement. The total strength of commercially pure aluminum is given as:

$$\sigma_y = \sigma_0 + \Delta\sigma_{gb} + M(\Delta\tau_{dis} + \Delta\tau_{ss}) \tag{7}$$

where σ_0 denotes the strength of annealed aluminum and $\Delta\sigma_{gb}$, $\Delta\tau_{ss}$, and $\Delta\tau_{dis}$ are the contributions due to the grain boundary strengthening, the solid solution hardening, and the dislocations hardening, respectively. The total dislocation density is the sum of the GND and SSD densities. At the center of the disk, the strain should be zero and hence the SSD density (ρ_{SSD})

is expected to be zero. However, the strain gradient is substantial and GNDs will be generated. The density of GNDs depends only on the strain gradient and the magnitude of Burger's vector (b) but is irrelevant to the alloying contents. In an idealized cylindrical coordinate, the strain gradient during HPT has only a radial component. Thus, the total amount of GNDs generated per unit volume is given by:

$$\rho_{GND,g} = \frac{1}{b}\frac{d\gamma}{dr} = \frac{2\pi Nb}{h} \tag{8}$$

The density of SSDs can be expressed by:

$$\rho_{SSD,g} = \varepsilon\left(\frac{K_A}{M\mu b\alpha}\right)^2 \tag{9}$$

where K_A is an alloy-dependent factor and μ denotes the shear modulus. The grain boundary strengthening is assumed to be inversely proportional to the average grain size:

$$\sigma_{gb} = \mu b\left(\frac{1}{D}\right) \tag{10}$$

The mean grain size, D, can be predicted assuming that the average grain boundary misorientation angle, θ, is determined by the dislocation density within the cell wall ρ_{GB}; α is a constant. This approach then provides the following relation:

$$D = 4.365\frac{\theta}{\rho_{GB}b} \tag{11}$$

Model predictions for Al-1050A alloy are given in **Table 3**. The modeling results show an excellent correspondence between measured and predicted average Vickers microhardness (determined as $H_V = \sigma_y/2.9$) at the center of Al-1050A samples processed by m-HPT for different turns. A key element of this model is the assumption that at very high strains, the dislocation density reaches a saturation value. However, the existing models are not capable to connect the flow stress with the observed microstructure evolution.

Turns	$\Delta\tau_{dis}$, **MPa**	$\Delta\sigma_{gb}$, **MPa**	σ_y, **MPa (predicted)**	H_V, **MPa (predicted)**	H_V, **MPa (measured)**
0	19	0.4	77	27	30
0.5	28	0.4	101	35	36
1	35	0.4	120	41	41
3	45	4	148	51	49
5	45	7	151	52	52
10	45	10	154	53	54

Table 3. Measured average Vickers microhardness (H_V) in the center of disk compared with model predictions for H_V and the different strengthening components [56].

Recently, Borodachenkova et al. [42] developed a microstructure-based model which allows more detailed analysis of the relative contributions of microstructure hardening mechanisms to the specific softening behavior in Al-30wt% Zn alloys during HPT. The experimental data indicated that the HPT processing leads to a strong softening process at the beginning of plastic deformation before the saturation stage is achieved. The softening is controlled by multiple mechanisms occurring simultaneously during the deformation, which are categorized in Borodachenkova et al. [42] as solid solution decomposition (τ_{ss}), Orowan mechanism (τ_{Orowan}), and dislocation strengthening (τ_{dis}). The total material strength is expressed as:

$$\tau - \tau_{ss} = \tau_{Orowan} + \tau_{dis} \tag{12}$$

Orowan mechanism refers to the strengthening due to the precipitates which act as impenetrable obstacles to the mobile dislocations. When a dislocation is pinned at Zn precipitates, it can still bow out between the precipitates and continue to glide if the driving stress is sufficient. The required stress for the bypass is related to the interspacing (ω) and size of the precipitates (d_p). In Al-Zn alloy, Zn precipitates are formed intensively in the Al grain and grain boundaries at the initial stage of deformation. With increasing strain, the Zn precipitates grow in size due to the diffusion of Zn atoms. Besides, the dislocations pinned at the Zn precipitates tend to transfer gradually into highly misoriented grain boundaries. At large strain, the bulk of Al grains is almost free of precipitates. In this case, theoretically τ_{Orowan} should increase rapidly at the beginning and vanish gradually. To express this process accurately, an empirical factor k_{or} is introduced in the classic Orowan mechanism law [57]:

$$\tau_{Orowan} = k_{or} \frac{0.85 \mu b ln\left(\frac{d_p}{b}\right)}{2\pi(\omega - d_p)} \tag{13}$$

k_{or} is imposed to be linearly increased to 1 and then reduced to 0 as a function of strain depending on the experimental observations.

The dislocation strengthening term τ_{dis} is expressed using the common Taylor law:

$$\tau_{dis} = \alpha \mu b \sqrt{\rho} \tag{14}$$

where α is the dislocation-dislocation interaction strength parameter. The dislocation density evolution is determined through the Kocks-Mecking-type equation. The dislocation mean free path is a complex matter since it is controlled initially by the precipitates interspacing and, after certain amount of strain, by the refined grain size. Its value is calculated through an empirical law as a function of the strain.

Solute strengthening is related to the interactions between dislocations and solute atoms. When a dislocation is traveling through a randomly distributed solute atom field (commonly known as Cottrell atmosphere), it will suffer a drag force induced by the solute atoms. For the dislocations pinned at obstacles, the diffusion of solute atoms into the dislocation core leads to an increase in the binding energy between the dislocations and their current location. The effect of solute strengthening depends on the concentration of solute atoms. In Al-Zn, the concentration of Zn atom keeps decreasing due to the precipitation process, which is the main reason for the softening phenomenon. In the work of Borodachenkova et al. [42], the approach of Mecking and Kocks [58] is modified to describe the solute strengthening:

$$\tau_{ss} = k_1\left(\tau_0 + a_0(c)\left(1 - \left(\frac{kT}{\Delta G} ln\frac{\dot{\gamma}_0}{\dot{\gamma}}\right)^{2/3}\right)\right) \tag{15}$$

where τ_0 denotes the lattice friction, ΔG is the activation energy, and a_0 is the value of thermal stress at 0 K, dependent on the Zn concentration in Al grains (c). a_0 can be written as [59]:

$$a_0 = \tau_p + \frac{3\Lambda}{2b^3}\left(\frac{\sqrt{2}\hat{U}^4}{A\omega_0}\right)^{1/3}\sqrt{c} \tag{16}$$

Here, τ_p is the Peierls stress, $\hat{U}$ the characteristic interaction energy between a single solute and a straight dislocation, ω_0 the characteristic range for the interaction, and A the line tension energy per unit length of dislocation. Comparison of the contributions of the hardening mechanisms is presented in **Figure 5**, which concludes that the shear stress is mostly controlled by the solid solution shear stress. The decomposition of super saturated solid solution plays a dominant role in the material properties of Al-30 wt% Zn alloy.

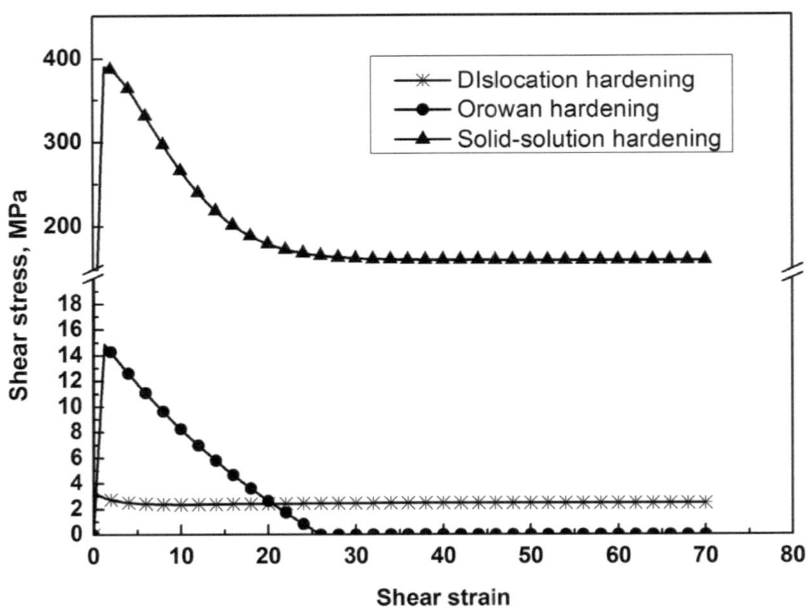

Figure 5. The comparison of the contribution of different hardening mechanisms predicted by the model of Borodachenkova et al. [42].

4.2. Finite-element approaches

The microstructure-based models described in the previous section consider the detailed evolution of the dislocation densities and other mechanisms such as the formation of second-phase precipitates. However, this type of models usually deals with the mechanical behavior at material point level. They cannot describe the heterogeneity of strain/microhardness distributions along the sample radius, which is quite evident during HPT. The finite-element method (FEM) shows an obvious advantage on this matter.

Despite the considerable interest in HPT technique, there are only very limited studies that focus on the heterogeneity of the plastic flow on the sample disk during the processing operation. According to the previous works, the FEM has been an effective tool to study the influence of the disk shape changes as well as the temperature evolutions occurring during HPT processing.

Figueiredo et al. [60] examined the quasi-constrained HPT processing with disks located within depressions on the inner anvil surfaces. The authors conducted the research using DEFORM-3D 10.0 software (Scientific Forming Technologies Corp., Columbus, OH) considering isothermal conditions. The following simulation conditions were taken into account: the applied pressures vary from 0.5 to 2.0 GPa, friction coefficients from 0 to 1 outside of the depressions, and torsional strains up to 1.5 turns. The simulation results show that the mean stresses vary linearly with the distance to the disk center. The authors reported that higher compressive stresses are observed in the disk center and lower stresses at the edge. The compressive mean stresses within the quasi-constrained volume decrease with the increasing extrusion of a ribbon of material between the anvils. The simulations indicate that the distribution of effective strains inside the quasi-constrained volume of the anvils is comparable to the prediction by ideal torsion according to Eqs. (2) and (3). The applied pressure and the friction coefficient outside the quasi-constrained volume play a minor role in the distribution of effective strain.

Later, Figueiredo et al. [61] studied the temperature distribution in quasi-constrained HPT. The calculation results show that the temperature increase within the sample is directly proportional to the material strength and the rotation speed. The temperature increasing rate varies almost linearly with the flow stress of the disk and seems to be independent of the material thermal properties. The study also indicates that a faster rotation speed leads to a higher deformation rate and consequently a higher rate of heating. However, it has been pointed out that the effect of the applied pressure in the HPT on the temperature increase is limited. Figueiredo et al. [61] predicted the evolution of the maximum temperature as a function of time in the iron disk. The results show that the temperature increases from 20 to 54°C during 600 s under 1-GPa pressure. When a much higher pressure of 16 GPa is applied, temperature varies from 20 to 62°C during 600 s. The increase in applied pressure leads to an increase in temperature due to the higher volume of material outflow between the anvils. The predicted temperature increase has been validated by using the experimental measurements of the temperature recorded in the upper anvil during HPT processing of Cu, Mo, and Al. The comparison of the calculated and the measured results depending on the applied pressure and the rotation speed is summarized in **Table 4**.

Material	τ (GPa)	ω (rpm)	P (GPa)	N=1	N=2	N=3	N=4
Cu	0.43	1	2	29.5 (26)	30.7 (28)	33.2 (31)	35.9 (33)
		0.5		25.3 (24)	26.2 (25)	27.9 (26)	29.0 (27)
		0.2		22.6 (21)	23.2 (22)	23.7 (23)	23.7 (23)
Mo	2.22	1	6	66.7 (51)	72.6 (65)	85.4 (87)	98.4 (100)
Al	0.21	5	5	45.7 (28)	47.9 (34)	51.3 (42)	55.5 (47)
		0.2		19.9 (20)	20.3 (20)	20.6 (21)	20.6 (22)

Table 4. Summary of material flow stress (estimated from hardness), the HPT processing parameters, and the temperature predicted at the workpiece (in °C). In parentheses, there are experimental measurements of the temperatures [61].

Song et al. [62] performed a more detailed analysis regarding the influence of the friction on the stress-strain distribution during HPT process. The results demonstrated that friction plays a more important role in the torsion stage than in the compression stage. This chapter also shows (see **Figure 6**) that the effect of the friction coefficient on the effective strain is more significant toward the disk-edge region. The variations of effective strain as a function of the friction coefficient for different locations in the workpiece are also shown in **Figure 6** (right). The authors declared that within the range of friction coefficient from 0.9 to 1.5, the effective strain increases sharply, particularly in the medium and edge areas.

The effect of the friction coefficient on effective strain distribution on the HPT sample is obtained in HPT-FEM simulations after 1 turn under 1-GPa pressure and 1 rpm rotation rate.

4.3. Coupling between micro-macro modeling and FEM

Lee et al. [63] embedded the dislocation density-based constitutive modeling a finite-element code to study the behaviors of pure copper during HPT. The coupling between FEM and microstructure-based constitutive model provides an excellent method for HPT-related

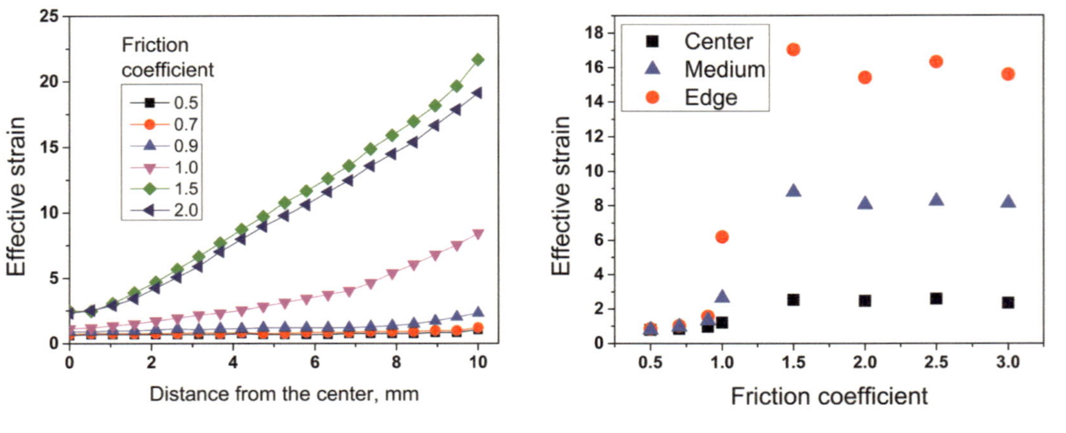

Figure 6. The effect of the friction coefficient on effective strain distribution on the HPT sample, obtained in HPT-FEM simulations after 1 turn under 1-GPa pressure and 1-rpm rotation rate.

predictions, which offers an adequate picture of the variation of the mechanistic parameters, such as stress, strain, and strain rate, as well as the evolution of the microstructural quantities, notably the dislocation density and the average grain size. In Lee et al. [63], the dislocation density evolution is described by the approach of Estrin [55]. The dislocation density-based constitutive model has been embedded in the rigid-plastic FEM package, DEFORM-3D ver. 6.1.

The stress and strain distribution during the HPT process has been analyzed, along with the dislocation density evolution and the concomitant variation of the dislocation cell size. The simulation results were compared with experimentally measured hardness and dislocation density (in the cell interiors and cell walls).The initial dislocation densities in the cell interior are 2.5×10^{13} and 5.0×10^{13} m^{-2}. The gradient of the dislocation density is observed along the sample diameter. After the compression stage, the dislocation densities increase and reach $7.27\times10^{14} m^{-2}$ in the center, 1.0×10^{15} m^{-2} in the middle, and 2.03×10^{15} m^{-2} at the edge region. After one anvil turn, the dislocation densities further increase, and the magnitude is 3.3×10^{15} m^{-2} in the center, 4.70×10^{15} m^{-2} in the middle, and 7.25×10^{15} m^{-2} at the edge region. The difference in the value of the dislocation density between the center and the edge has been explained by the fact that the torsional strain is proportional to the distance from the center as given in Eq. (2). The authors also compared the dislocation density predicted by FEM with the experimental data synchrotron X-ray powder diffraction (XRD) analysis, and a good agreement is achieved.

Author details

Marina Borodachenkova[1]*, Wei Wen[2] and António Manuel de Bastos Pereira[1]

*Address all correspondence to: m.borodachenkova@tue.nl

1 Center for Mechanical Technology and Automation, Mechanical Engineering Department, University of Aveiro, Aveiro, Portugal

2 Materials Science and Technology Division, Los Alamos National Laboratory, Los Alamos, New Mexico, USA

References

[1] Meyers MA, Mishra A, Benson DJ. Mechanical properties of nanocrystalline materials. Progress in Materials Science. 2006;**51**:427–556

[2] Mazilkin AA, Straumal BB, Rabkin E, Baretzky B, Enders S, Protasova SG, Kogtenkova OA, Valiev RZ. Softening of nanostructured Al-Zn and Al-Mg alloys after severe plastic deformation. Acta Materialia. 2006;**54**:3933–3939

[3] Zhu YT, Lowe TC, Langdon TG. Performance and applications of nanostructured materials produced by severe plastic deformation. Scripta Materialia. 2004;**51**:825–830

[4] Zhilyaev AP, Lee S, Nurislamova GV, Valiev RZ, Langdon TG. Microhardness and microstructural evolution in pure nickel during high-pressure torsion. Scripta Materialia. 2001;**44**:2753–2758

[5] Islamgaliev RK, Chmelik F, Kuzel R. Thermal structure changes in copper and nickel processed by severe plastic deformation. Materials Science and Engineering: A. 1997;**234-236**:335–338

[6] Zhilyaev AP, McNelley TR, Langdon TG. Evolution of microstructure and microtexture in fcc metals during high-pressure torsion. Journal of Materials Science. 2007;**42**:1517–1528

[7] Figueiredo RB, Langdon TG. Using severe plastic deformation for the processing of advanced engineering materials. Materials Transactions. 2009;**50**:1613–1619

[8] Krystian M, Setman D, Mingler B, Krexner G, Zehetbauer MJ. Formation of superabundant vacancies in nano-Pd-H generated by high-pressure torsion. Scripta Materialia. 2010;**62**:49–52

[9] Zhang J, Gao N, Starink MJ. Al-Mg-Cu based alloys and pure Al processed by high pressure torsion: The influence of alloying additions on strengthening. Materials Science and Engineering: A. 2010;**527**:3472–3479

[10] Jiang H, Zhu YT, Butt DP, Alexandrov IV, Lowe TC. Microstructural evolution, microhardness and thermal stability of HPT-processed Cu. Materials Science and Engineering: A. 2000;**290**:128–138

[11] Vorhauer A, Pippan R. On the homogeneity of deformation by high pressure torsion. Scripta Materialia. 2004;**51**:921–925

[12] Yang Z, Welzel U. Microstructure-microhardness relation of nanostructured Ni produced by high-pressure torsion. Materials Letters. 2005;**59**:3406–3409

[13] Zhilyaev AP, Nurislamova GV, Kim BK, Baró MD, Szpunar JA, Langdon TG. Experimental parameters influencing grain refinement and microstructural evolution during high-pressure torsion. Acta Materialia. 2003;**51**:753–765

[14] Xu C, Horita Z, Langdon TG. The evolution of homogeneity in processing by high-pressure torsion. Acta Materialia. 2007;**55**:203–212

[15] Kawasaki M, Alhajeri SN, Xu C, Langdon TG. The development of hardness homogeneity in pure aluminum and aluminum alloy disks processed by high-pressure torsion. Materials Science and Engineering: A. 2011;**529**:345–351

[16] An XH, Wu SD, Zhang ZF, Figueiredo RB, Gao N, Langdon TG. Evolution of microstructural homogeneity in copper processed by high-pressure torsion. Scripta Materialia. 2010;**63**:560–563

[17] Zhilyaev AP, Oh-ishi K, Langdon TG, McNelley TR. Microstructural evolution in commercial purity aluminum during high-pressure torsion. Materials Science and Engineering: A. 2005;**410-411**:277–280

[18] Mishra RS, Valiev RZ, McFadden SX, Mukherjee AK. Tensile superplasticity in a nanocrystalline nickel aluminide. Materials Science and Engineering: A. 1998;**252**:174–178

[19] Sakai G, Horita Z, Langdon TG. Grain refinement and superplasticity in an aluminum alloy processed by high-pressure torsion. Materials Science and Engineering: A. 2005;**393**:344–351

[20] Lugo N, Llorca N, Cabrera JM, Horita Z. Microstructures and mechanical properties of pure copper deformed severely by equal-channel angular pressing and high pressure torsion. Materials Science and Engineering: A. 2008;**477**:366–371

[21] Edalati K, Fujioka T, Horita Z. Microstructure and mechanical properties of pure Cu processed by high-pressure torsion. Materials Science and Engineering: A. 2008;**497**:168–173

[22] Edalati K, Horita Z, Yagi S, Matsubara E. Allotropic phase transformation of pure zirconium by high-pressure torsion. Materials Science and Engineering: A. 2009;**523**:277–281

[23] Todaka Y, Sasaki J, Moto T, Umemoto M. Bulk submicrocrystalline ω-Ti produced by high-pressure torsion straining. Scripta Materialia. 2008;**59**:615–618

[24] Ito Y, Horita Z. Microstructural evolution in pure aluminum processed by high-pressure torsion. Materials Science and Engineering: A. 2009;**503**:32–36

[25] Edalati K, Horita Z. Significance of homologous temperature in softening behavior and grain size of pure metals processed by high-pressure torsion. Materials Science and Engineering: A. 2011;**528**:7514–7523

[26] Edalati K, Yamamoto A, Horita Z, Ishihara T. High-pressure torsion of pure magnesium: Evolution of mechanical properties, microstructures and hydrogen storage capacity with equivalent strain. Scripta Materialia. 2011;**64**:880–883

[27] Ni S, Wang YB, Liao XZ, Alhajeri SN, Li HQ, Zhao YH, Lavernia EJ, Ringer SP, Langdon TG, Zhu YT. Strain hardening and softening in a nanocrystalline Ni–Fe alloy induced by severe plastic deformation. Strain. 2011;**528**:3398–3403

[28] Srinivasarao B, Zhilyaev AP, Langdon TG, Pérez-Prado MT. On the relation between the microstructure and the mechanical behavior of pure Zn processed by high pressure torsion. Materials Science and Engineering: A. 2013;**562**:196–202

[29] Liddicoat PV, Liao X-Z, Zhao Y, Zhu Y, Murashkin MY, Lavernia EJ, Valiev RZ, Ringer SP. Nanostructural hierarchy increases the strength of aluminium alloys. Nature Communications. 2010;**1**:63

[30] Dutkiewicz J, Kuśnierz J, Maziarz W, Lejkowska M, Garbacz H, Lewandowska M, Dobromyslov AV, Kurzydłowski KJ. Microstructure and mechanical properties of nanocrystalline titanium and Ti-Ta-Nb alloy manufactured using various deformation methods. Physica Status Solidi: Applications and Materials Science. 2005;**202**:2309–2320

[31] Sabbaghianrad S, Kawasaki M, Langdon TG. Microstructural evolution and the mechanical properties of an aluminum alloy processed by high-pressure torsion. Journal of Materials Science. 2012;**47**:7789–7795

[32] Sabbaghianrad S., Wongsa-Ngam J, Kawasaki M, Langdon TG. An examination of the saturation microstructures achieved in ultrafine-grained metals processed by high-pressure torsion. Journal of Materials Research and Technology. 2014;**3**:319–326

[33] Huang Y, Xu J, Langdon TG. Developing ultrafine-grained materials with high strength and good ductility for micro-forming applications. MATEC Web of Conferences. 2015;21:7002

[34] Xu J, Li J, Wang CT, Shan D, Guo B, Langdon TG. Evidence for an early softening behavior in pure copper processed by high-pressure torsion. Journal of Materials Science. 2016;**51**:1923–1930

[35] Stolyarov VV, Soshnikova EP, Brodova IG, Bashlykov DV, Kil'mametov AR. The aging effect in an ultrafine-grained Al-5% Fe alloy produced by severe plastic deformation. Physics of Metals and Metallography. 2002;**93**:567–574

[36] Valiev RZ, Ivanisenko YV, Rauch EF, Baudelet B. Structure and deformaton behaviour of Armco iron subjected to severe plastic deformation. Acta Materialia. 1996;**44**:4705–4712

[37] Stráská J, Janeček M, Gubicza J, Krajňák T, Yoon EY, Kim HS. Evolution of microstructure and hardness in AZ31 alloy processed by high pressure torsion. Materials Science and Engineering: A. 2015;**625**:98–106

[38] Maury N, Zhang NX, Huang Y, Zhilyaev AP, Langdon TG. A critical examination of pure tantalum processed by high-pressure torsion. Materials Science and Engineering: A. 2015;**638**:174–182

[39] Podolskiy AV, Mangler C, Schafler E, Tabachnikova ED, Zehetbauer MJ. Microstructure and mechanical properties of high purity nanostructured titanium processed by high pressure torsion at temperatures 300 and 77 K. Journal of Materials Science. 2013;**48**:4689–4697

[40] Edalati K, Horita Z, Mine Y. High-pressure torsion of hafnium. Materials Science and Engineering: A. 2010;**527**;2136–2141

[41] Isik M, Niinomi M, Liu H, Cho K, Nakai M, Horita Z, Sato S, Narushima T, Yilmazer H, Nagasako M. Grain refinement mechanism and evolution of dislocation structure of Co-Cr-Mo alloy subjected to high-pressure torsion. Materials Transactions. 2016;**57**:1109–1118

[42] Borodachenkova M, Barlat F, Wen W, Bastos A, Grácio JJ. A microstructure-based model for describing the material properties of Al-Zn alloys during high pressure torsion. International Journal of Plasticity. 2015;**68**:150–163

[43] Liu M, Roven HJ, Liu X, Murashkin M, Valiev RZ, Ungár T, Balogh L. Grain refinement in nanostructured Al-Mg alloys subjected to high pressure torsion. Journal of Materials Science. 2010;**45**:4659–4664

[44] Ito Y, Edalati K, Horita Z. High-pressure torsion of aluminum with ultrahigh purity (99.9999%) and occurrence of inverse Hall-Petch relationship. Materials Science and Engineering: A. 2017;**679**:428–434

[45] Pang Y, Li P, Kim HS, Gong Y, Shen Y, Sun L, Zhu X. Strain softening induced by high pressure torsion in copper alloys. Materials Transactions. 2015;**56**:1658–1663

[46] Mazilkin AA, Straumal BB, Borodachenkova MV, Valiev RZ, Kogtenkova OA, Baretzky B. Gradual softening of Al-Zn alloys during high-pressure torsion. Materials Letters. 2012;**84**:63–65

[47] Kawasaki M, Langdon TG. Review: Achieving superplasticity in metals processed by high-pressure torsion. Journal of Materials Science. 2014;**49**:6487–6496

[48] Ashida M, Chen P, Doi H, Tsutsumi Y, Hanawa T, Horita Z. Superplasticity in the Ti-6Al-7Nb alloy processed by high-pressure torsion. Materials Science and Engineering: A. 2015;**640**:449–453

[49] Matsunoshita H, Edalati K, Furui M, Horita Z. Ultrafine-grained magnesium-lithium alloy processed by high-pressure torsion: Low-temperature superplasticity and potential for hydroforming. Materials Science & Engineering A. 2015;**640**:443–448

[50] Alhamidi A, Horita Z. Grain refinement and high strain rate superplasticity in aluminium 2024 alloy processed by high-pressure torsion. Materials Science and Engineering: A. 2015;**622**:139–145

[51] Al-Zubaydi ASJ, Zhilyaev AP, Wang SC, Reed PAS. Superplastic behaviour of AZ91 magnesium alloy processed by high-pressure torsion. Materials Science and Engineering: A. 2015;**637**:1–11

[52] Sabbaghianrad S, Langdon TG. Developing superplasticity in an aluminum matrix composite processed by high-pressure torsion. Materials Science and Engineering: A. 2016;**655**:36–43

[53] Fu J, Ding H, Huang Y, Pereira PHR, Zhang W, Langdon TG. Grain refining of a Ti-6Al-4V alloy by high-pressure torsion and low temperature superplasticity. Materials Letters. 2015;**5**:281–286

[54] Kocks UF, Mecking H. Physics and phenomenology of strain hardening: The FCC case. Progress in Materials Science. 2003;**48**:171–273

[55] Estrin Y. Dislocation theory based constitutive modelling: Foundations and applications. Journal of Materials Processing Technology. 1998;**80-81**:33–39

[56] Zhang J, Gao N, Starink MJ. Microstructure development and hardening during high pressure torsion of commercially pure aluminium: Strain reversal experiments and a dislocation based model. Materials Science and Engineering: A. 2011;**528**:2581–2591

[57] Kocks UF. A statistical theory of flow stress and work-hardening. Philosophical Magazine. 1966;**13**:541–566

[58] Mecking H, Kocks UF. Kinetics of flow and strain-hardening. Acta Metallurgica. 1981;**29**: 1865–1875

[59] Olmsted DL, Hector LG, Curtin WA. Molecular dynamics study of solute strengthening in Al/Mg alloys. Journal of the Mechanics and Physics of Solids. 2006;**54**:1763–1788

[60] Figueiredo RB, Cetlin PR, Langdon TG. Using finite element modeling to examine the flow processes in quasi-constrained high-pressure torsion. Materials Science and Engineering: A. 2011;**528**:8198–8204

[61] Figueiredo RB, Pereira PHR, Aguilar MTP, Cetlin PR, Langdon TG. Using finite element modeling to examine the temperature distribution in quasi-constrained high-pressure torsion. Acta Materialia. 2012;**60**:3190–3198

[62] Song Y, Wang W, Gao D, Yoon EY, Lee DJ, Kim HS. Finite element analysis of the effect of friction in high pressure torsion. Metals and Materials International. 2014;**20**:445–450

[63] Lee DJ, Yoon EY, Ahn D-H, Park BH, Park HW, Park LJ, Estrin Y, Kim HS. Dislocation density-based finite element analysis of large strain deformation behavior of copper under high-pressure torsion. Acta Materialia. 2014;**76**:281–293

Section 3

SPD: Novelty and application examples

Chapter 5

Cu-Cr and Cu-Fe Alloys Processed by New Severe Plastic Deformation: Microstructure and Properties

Kinga Rodak

Additional information is available at the end of the chapter

http://dx.doi.org/10.5772/intechopen.68954

Abstract

In this chapter, two techniques have been proposed for grain refinement in Cu-Cr and Cu-Fe alloys in different heat treatment conditions. First method, known as rolling with cyclic movement of rolls (RCMR), is appropriate for the manufacturing of ultrafine grained sheets and plates. The second method is called compression with oscillatory torsion (COT). Structural investigations of alloys were carried out, in particular, using a cold field emission gun/scanning electron microscope (FEG/SEM) equipped with an electron backscattering diffraction (EBSD) detector and a scanning transmission electron microscope (STEM). Quantitative studies of the microstructure based on the STEM images were performed using the "MET-ILO" software package. Mechanical properties were determined using an MST QTest/10 instrument equipped with digital image correlation (DIC). Based on the SEM and STEM observations, it has been shown that the alloys may exhibit a refinement of the ultra fine grained (UFG) structure in the 200–500 nm range with a mixture of low- and high-angle boundaries. Although the microstructure was refined significantly, the heterogeneity of the microstructure after the application of a high total effective strain is observed. Moreover, the low-angle boundaries formed at the early stages do not continuously transform into high-angle boundaries.

Keywords: SPD, mechanical properties, microstructure, STEM, EBSD

1. Introduction

Severe plastic deformation (SPD) is a metal forming process used to introduce ultra-high plastic strains into a bulk metal in order to refine the grain structure of metallic materials to the submicrometer (100–1000 nm) or even nanometer (less than 100 nm) range. A finer grain size increases the hardness, the yield stress, and the fracture toughness of the material [1–9]. Overall, SPD deformation is recognized as a potential tool for superplastic deformation at lower temperatures and high strain rates [2–5]. While in the conventional metal forming processes, the imposed plastic strain is generally low, SPD imposes an extremely large strain on the bulk metal without changing the shape of the metal necessary for chosen techniques [2, 3]. Practically, obtaining a large plastic deformation is difficult because in most metal-forming processes the deformation is limited by the failure of either the material or the tool. Considering the deformation processing conditions, the heterogeneity in microstructure formation was often observed across the bulk specimen depending on the introduced strain [3, 6, 7]. From the literature it is known [5–7] that very fine and highly disoriented grain structures obtained in different bulk metals and alloys are created as a result of short- and long-range intersecting shear bands produced by plastic deformation. Additionally, the local dynamic recovery and recrystallization processes [5, 7] also contribute to the grain refinement. Distinct structures are either dislocation-free or with dislocations and fine grains that are highly or weakly disoriented and are obtained during the SPD deformation. The evolution and characterization of the new grain/subgrain boundaries appear to be very important for their influence on the mechanical and functional properties of SPD materials. Additionally, the SPD problem is connected with the effect of the strain path on the structure formation, i.e. a broad variety of structures that show differences in the grain size, shape, and crystallographic texture [7–9].

The influence of SPD on the microstructure refinement is more complicated when the precipitation strengthening mechanism of alloys is taken into account. The effect of the secondary phase particles on the grain refinement mechanisms during severe deformation processing is still topical and has been explored extensively [10–14]. The presence of second phase particles is a source of a significant strengthening resulting from the interaction of dislocations. The strength of this interaction depends on the chemical composition, size, distribution, and the degree of coherence with the matrix. In engineering alloys, the second phase particles may take the form of dispersoids, fine precipitates, and large primary particles that may show quite different effects on the grain refinement during SPD processing.

Two original methods patented at Silesian University of Technology, Faculty of Materials Engineering and Metallurgy in Poland for grain refinement are discussed in this paper: the first method is called rolling with cyclic movement of rolls (RCMR) and is appropriate for the manufacturing of ultrafine grained sheets and plates. The second method allows large deformations and is called compression with oscillatory torsion (COT). This process combines two deformation methods: (1) the compression process which is effective for obtaining high mechanical properties and a microstructure composed of elongated grains with a high dislocation density; (2) the torsion process which is effective for obtaining a higher plasticity and a microstructure consisting of quite equiaxed subgrains with a small dislocation density inside the subgrains. The combination of these methods in a single deformation procedure enables the optimization of the microstructure and mechanical properties. COT investigations were performed for Cu and Al [14–19] and this method was recognized as an effective tool for obtaining ultrafine grains/subgrains with a mixture of low- and high-angle grain boundaries.

In this chapter, we demonstrate the effectiveness of COT and RCMR deformation as a technique for studying large strain deformation in Cu-Fe and Cu-Cr alloys at room temperature.

Cu-Cr alloys are used in numerous applications where a combination of excellent mechanical strength and electrical conductivity is required [20–23]. They can be important materials for railway contact wires and electrodes for spot welding. Meanwhile, Cu-Fe alloys are commonly used as electrical device components such as semiconductor lead frames and electrical connectors [24].

The Cu-Cr and Cu-Fe alloys belong to the class of precipitation-hardened alloys, and therefore, the problem of the initial structure (after solution and aging treatment at different parameters) is discussed in this paper. The aim of the present research is also extended to the evaluation of whether it is possible to obtain smaller grain sizes and, consequently, to improve the mechanical properties of the material if the processed materials have different initial structures (different states of heat treatment). The obtained results may be useful for constructing a complete picture of the structure and properties evolution in these alloys during RCMR and COT processing.

It should be noted that Cu-Cr and Cu-Fe alloys were chosen for the present study because these alloys are readily produced by SPD techniques at room temperature. This is because these alloys belong to the class of face-centered cubic precipitation hardenable alloys that are much more thermally and mechanically stable during the deformation than the pure Cu without the structural instabilities such as grain coarsening.

2. Refinement structure of Cu-Cr and Cu-Fe alloys by SPD techniques

Annealed high-purity of Cu exhibits a low strength of nearly 100–200 MPa and attractive physical properties such as its high electrical conductivity. The extensive dynamic recovery balancing the multiplications and annihilation of the dislocations is the limiting factor for the grain refinement in Cu [15, 16].

The introduction of solute atoms in the Cu matrix is the first route for the strengthening of conventional Cu alloys. The introduction of solute atoms into a solid solution produces an alloy that is stronger than the pure Cu due to the differences in the radius, modulus, and valence between the Cu matrix and solute atoms [20–24]. On the other hand, the alloying process results in the degradation of electrical conductivity. Moreover, after the deformation (SPD), the strength evidently increases but the conductivity decreases slightly due to the introduction of a large quantity of dislocations by cold deformation. The selection of the optimum properties for electrical applications always involves a trade-off between the mechanical and electrical properties. The precipitation effect during age hardening in Cu alloys gives rise to a subsequent effect improving the strength and also the electrical properties. The presence of precipitations reduces the dislocation mobility and the rate of dynamic recovery in alloys. In this case, the dislocations in the deformed Cu cannot be easily annihilated, very often leading to nonequilibrium dislocation clustering near the boundaries.

Little data is available in the literature for the Cu-Cr and Cu-Fe alloys after the application of SPD. The available data for the influence of SPD on the structure and mechanical properties are presented and summarized in **Table 1**. Examination of the data clearly shows that SPD processing was applied for the alloy after casting or quenching. It should be noted that an application of SPD deformation for the refinement of the structure of these alloys can produce grain sizes in the 100–400 nm range.

Material/SPD process	Initial state	Microstructure	Mechanical properties	References
Cu-0.5Cr/ECAP route A, $\varepsilon = 6.4$	Solid solution at 1000°C/0.5 h	Pancake shaped grains with high-angle boundaries with thickness of 100 nm. Many boundaries are curved, indicating the presence of high internal stress	UTS= 484 MPa Electrical cond. 35% of IACS HV0.3 = 145	[23]
Cu-0.36Cr ECAP route A,	Solid solution at 1025°C/40 min	Pancake-fragmented structure with lamellar boundaries. Grain diameter of 0.41 μm. Fraction of HABs0.75	HV0.2 = 150	[21]
Cu-0.75Cr, HPT, $\varepsilon = 4.8$	Solid solution at 1025°C/2 h	Average grain size-209 nm, dislocation density $\rho = 38 \times 10^{14}\ m^{-2}$	HV (MPa) = 1740 after HPT Electrical cond. 34% of IACS	[20]
Cu-0.5Cr-0.1Ag HPT, 6 GPa, 10 rev.	Solid solution at 1025°C/1 h	Homogeneous structure UFG with the mean grain size of about 200 nm	UTS = 840 MPa, *A* = 10%	[22]
Cu-Fe_P ECAP route Bc route A	As-cast alloy	Ultrafine grain structure with most dislocations found on the grain boundaries and the interior of the grain showing a lower dislocation density. The grains are not far from equiaxed, ranging from 50 to larger than 200 nm	HV = 166	[24]

Table 1. Effect of SPD deformation on grain size refinement in Cu alloys.

3. Materials and experiment

Copper alloys with addition of 0.6 wt% Cr (C18200) and 2 wt% Fe (C19400) were produced using induction melting of highly pure components. **Figure 1** shows the microstructure of the samples after casting with columnar grains. Inside the grains of CuCr0.6 alloy, for example, the Cu, Cr, and P elements are homogeneously distributed in the alloy after the casting (**Figure 2**). The ingots with the diameter of 50 mm were hot-deformed on the rods. Subsequently, the rolled bars underwent different heat treatments such as the solution treatment at 1000°C for 3 h (P state), followed by quenching into iced water and aging at 500°C for 2 h (S1 state), and 700°C for 24 h (S2 state).The microstructures of the CuCr0.6 and CuFe2 alloys after the solution treatment are characterized by the presence of equiaxed grains with heterogeneously distributed undissolved large Fe and Cr precipitates [14, 22, 23]. The aging treatment at 500°C for 2 h results in the formation of homogeneously distributed coherent precipitates within the matrix. As the aging temperature and time are increased, the precipitates lost coherence with

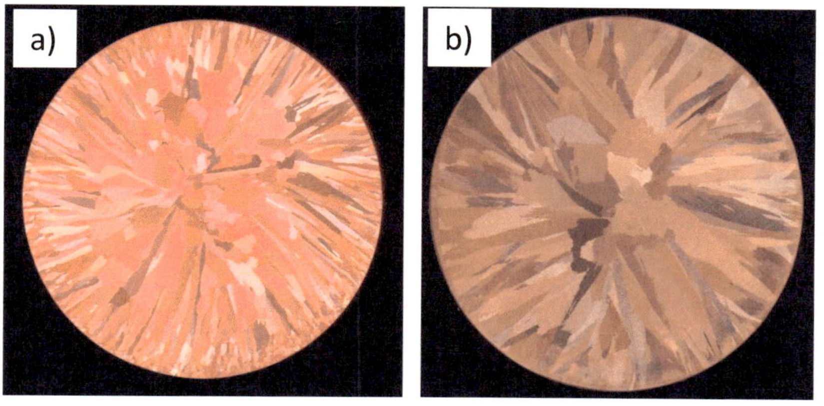

Figure 1. Microstructure of CuFe2 (a) and CuCr0.6 (b) samples after casting.

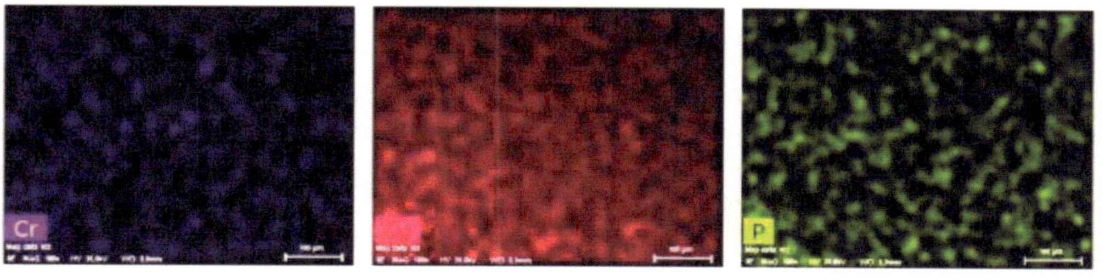

Figure 2. Distribution of Cu, Cr, and Pin the CuCr0.6 alloy after casting.

the matrix and the number of the particles within the matrix decreased (the spacing between the particles increased). The measured microstructural parameters of the precipitates are presented in **Table 2**. The samples were mechanically machined for appropriate dimensions for RCMR and COT deformation.

Scanning electron microscopy (FEG SEM INSPECT F by FEI equipped with an electron backscattering diffraction (EBSD) detector) was used to characterize the microstructures of the deformed alloys. Additionally, a scanning transmission electron microscope (STEM) Hitachi HD-2300A operated at 200 kV was applied for substructure characterization. The quantitative

State of material	CuCr0.6		CuFe2	
	d	*λ*	*d*	*λ*
P	0.7 μm	2.5 μm	0.5 μm	4.5 μm
S1	20 nm	60 nm	20 nm	40 nm
S2	270 nm	520 nm	100 nm	200 nm

Table 2. Measured microstructural parameters: *d*: average particle diameter, *λ*: average distance between the particles in CuCr0.6 and CuFe2 alloys after heat treatment.

studies of the structure parameters (for example: grain/subgrain and precipitate sizes) based on the STEM images were performed using the "MET-ILO" software package.

The mechanical properties were determined using an MST QTest/10 instrument equipped with digital image correlation (DIC). The use of the DIC method is advantageous due to its non-contact character and ability to perform high precision strain measurements. A SIGMATEST electric conductivity instrument was used to measure the conductivity. Due to the heterogeneity of the plastic deformation in the sample after the COT deformation, structural studies and mechanical investigations were performed on the samples extracted at the distance of 0.8 of the radius in the longitudinal plane section. The heterogeneity of the plastic deformation in the RCMR method causes a considerable differentiation of the structure. Microstructural observations (SEM) and evaluation of the mechanical properties were performed in the transverse plane section located at the height of ~0.8 of the specimen height. Since STEM analysis was not possible on the transverse section due to the small dimensions of the sample, STEM observations of the thin foils parallel to the rolling plane were performed at the distance of 0.6 of the specimen height.

Vickers hardness (HV0.2) measurements were carried out using a FM-310 Future-Tech hardness machine with the load of 200 g for 15 s. Microhardness measurements were performed in a plane parallel to the compression direction. To accurately describe the heterogeneity occurring during the COT processing, hardness maps were obtained for the longitudinal sections of the samples. The distance between the measuring points was about 0.5 mm, giving approximately 200 measurement points used to create the hardness maps.

The mechanical properties were determined using an MST QTest/10 machine equipped with digital image correlation (DIC). The tensile tests were performed at room temperature at the initial strain rate of 1×10^{-3} s^{-1}. Small tensile specimens with the total length and thickness of 8.6 and 0.3 mm, respectively, were used to measure the mechanical properties.

4. Production of ultrafine grained structure of CuCr0.6 and CuFe2 alloys by RCMR method

Several excellent techniques have been developed for creating tapes and strips products including the methods of constrained groove pressing (CGP) and accumulative roll-bonding (ARB). According to the reports in the literature [2], the CGP and ARB techniques exhibit several advantages over other SPD processes because (1) they do not require forming facilities with a large load capacity and expensive dies; (2) the amount of the material that can be produced is not limited. These methods are appropriate for the manufacturing of nanocrystalline and ultrafine grained sheets and plates. Rolling with cyclic movement of rolls method (RCMR) is a severe plastic deformation process that allows large deformations and is based on the rolling connected with the movement of the material layers in a direction perpendicular to the main direction of the rolling. By repeating this procedure, very high strains have been introduced into the material and a significant structure refining effect is obtained. This original method of deformation has been patented by Silesian University of Technology,

Faculty of Materials Engineering and Metallurgy in Poland. Compared to the various other SPD methods, the RCMR method exhibits the following advantages: the method is simple and can be carried out using conventional testing machines. It is possible to obtain a wide range of strain rates and high total effective strain values. RCMR can be applied to large-scale workplaces and thus has the potential for application in industry.

Figure 3 shows the RCMR setup. The rolling mill consists of two working rolls, the power unit, and the mechanism for the cyclic movement of the rolls transverse to the rolling direction. (The rolling mill without the mechanism for the cyclic movements of rolls is a typical example of setup for conventional cold-rolling.) During RCMR processing, the rolls rotate around an axis and, in addition, axial movements of the rolls in opposite directions are realized.

The structure and mechanical properties are found to depend strongly on the imposed total effective strain ε_{ft} given by Eqs. (1)–(3):

$$\varepsilon_{ft} = \sum_{i=1}^{n} \sqrt{\varepsilon_{hi}^2 + \varepsilon_{ti}^2}, \tag{1}$$

$$\varepsilon_{hi} = /\ln \frac{h_i}{h_{i-1}}/, \tag{2}$$

$$\varepsilon_{ti} = \frac{4 \cdot f \cdot A \cdot \sqrt{(h_i - h_{i-1}) \cdot \frac{D}{2}}}{\sqrt{3} \cdot v \cdot (h_{i-1} + h_i)} \tag{3}$$

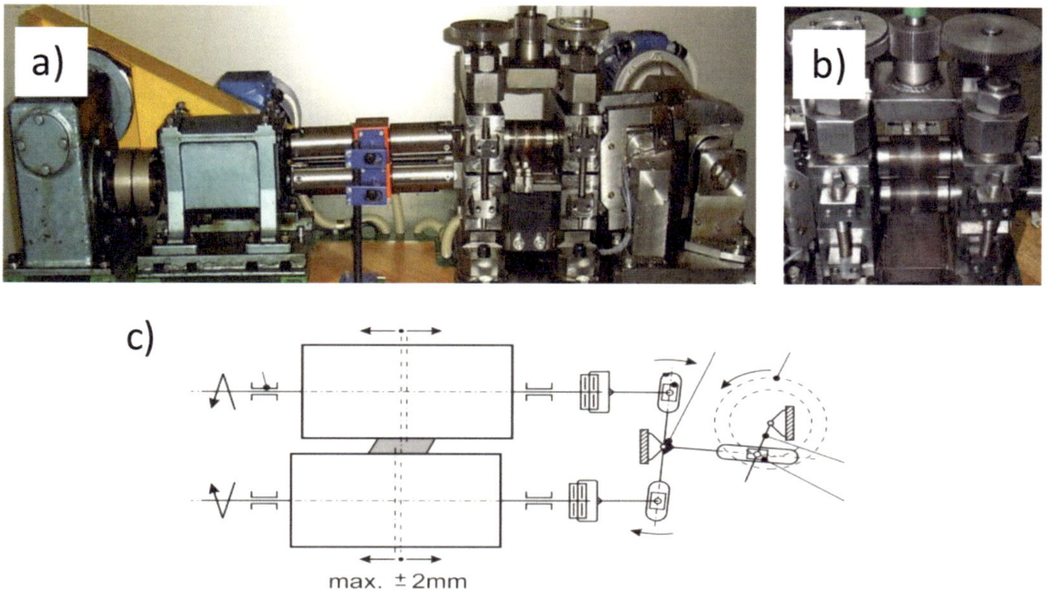

Figure 3. (a) RCMR setup; (b) working rolls; and (c) RCMR scheme.

Here, εft is the total effective strain, εhi is the strain contributed by the rolling reduction, εti is the strain included by the transverse movement of the working rolls, n is the number of passes, hi−1, hi are the heights of the sample before and after the unit pass (reduction), and A is the amplitude of the cyclic roll movement, v – (%), rolling rate , f – frequency of the transverse roll movement, D- diameter of rolls – here D=100mm

The deformation path can be controlled by changing the proportions of the following parameters: the rolling reduction ε_h (%), rolling rate v (rpm), amplitude of the transverse roll movement A (mm), and frequency of the transverse roll movement f (Hz).

The changes in the rolling forces and torques with the number of passes for the CuCr0.6 alloy are presented in **Figure 4**.

For all passes, the values of the rolling forces and rolling torques are larger for the conventional rolling than for RCMR. Examination of the obtained data shows that the additional transverse deformation during RCMR has a significant effect, decreasing the processing parameters. The data for the height of the samples after the deformation (1–6 passes) for rolling and RCMR process are

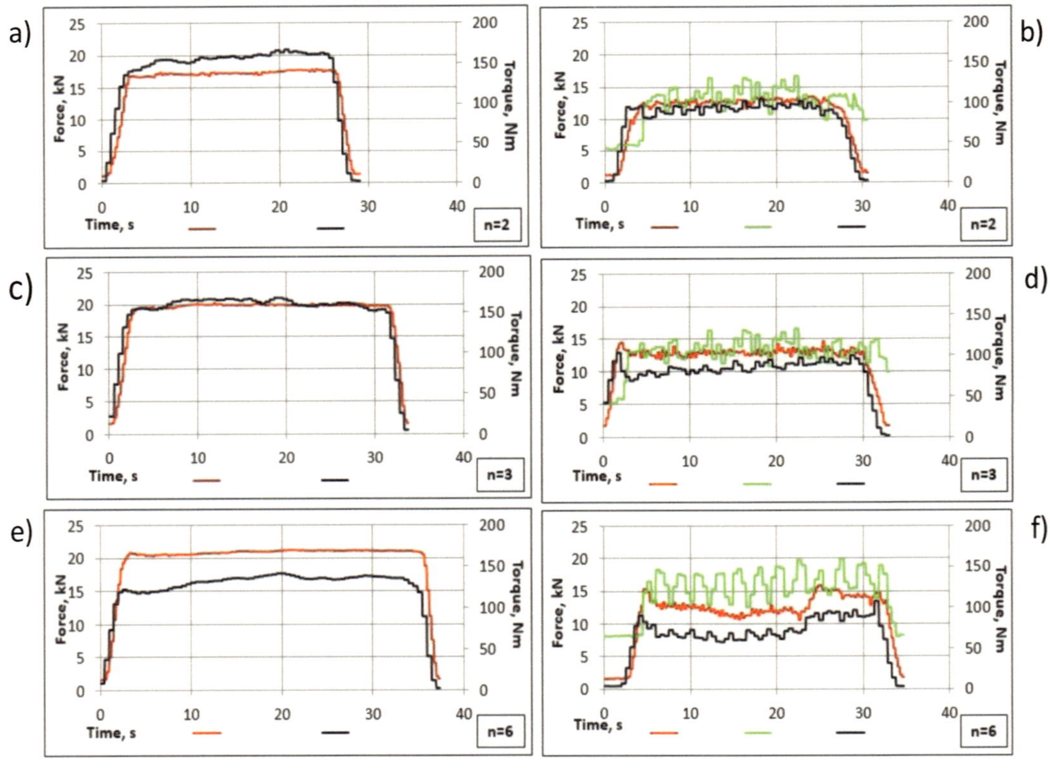

Figure 4. Changes in rolling forces and torques with the number of passes for CuCr0.6 alloy sample (S2 state), (a), (c), (e) conventional rolling, (b), (d), (f) RCMR. Parameters: rolling rate $v = 1$ (rpm), amplitude of the transverse roll movement $A = 0.8$ (mm), and frequency of the transverse roll movement $f = 1.5$ (Hz). Legend: red line-rolling force, black line-torque, green line-axial force.

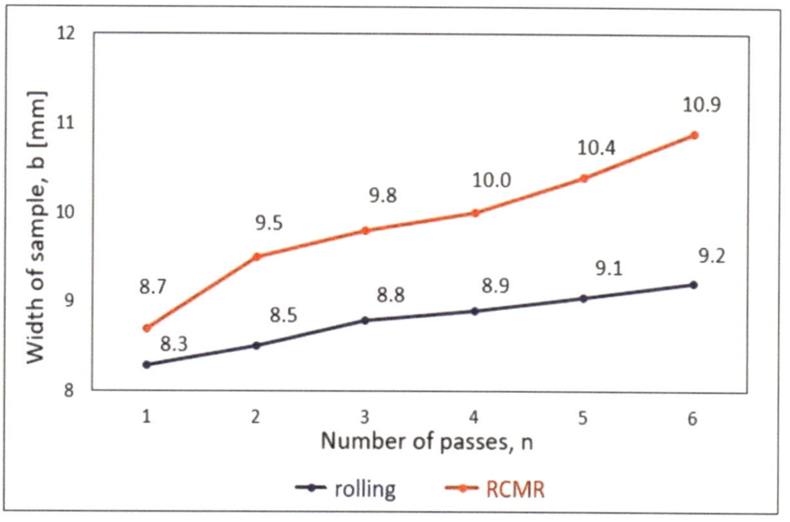

Figure 5. Changes of sample width of CuFe2 (P state) for rolling and RCMR as a function of the number of passes. Parameters: rolling rate v = 0.7 (rpm), amplitude of the transverse roll movement A = 0.8 (mm), and frequency of the transverse roll movement f = 1.5 (Hz).

presented in **Figure 5**. These results show that larger heights are obtained for the rolling samples than for the RCMR samples. However, the obtained values are very close to each other.

The changes in the total effective strain as a function of the number of passes during rolling and RCMR of the CuCr0.6 alloy at selected parameters are presented in **Figure 6** while **Figure 7** presents the obtained results for the dependence of the total effective strain of the CuFe2 alloy on the rolling rate. The obtained results show that the rate of the RCMR has a significant impact on the value of the total effective strain. Generally, examination of the obtained results shows that for the RCMR process, the total effective strain values in each pass are different and for RCMR, the values are 3–5 times higher than those obtained for conventional rolling. This means that the calculated total effective strain values in the RCMR process are not constant during subsequent passes. This is due to the heterogeneity in the deformation in a volume of the rolled strip and the phenomena occurring at the contact surfaces between the working rolls and strips. The obtained results show evidently that in the conventional metal forming process such as cold rolling, the imposed plastic strain is generally less than about 2.0 (**Figure 6**). In the RCMR process an extremely large strain is possible to impose without additional changes in shape (**Figures 5** and **6**).

The sample temperature during conventional deformation and RCMR was investigated using a thermal imaging camera (**Figure 8a** and **b**). The temperature rise was calculated as the difference between the maximum of the surface temperature of the rolled sample and the ambient temperature (adopted as 24°C) (**Figure 8c**). It is evident that for the RCMR samples, the temperature rise is higher than that for the conventional sample. Increase of the temperature during RCMR is due to the additional work done by the plastic deformation resulting from the transverse roll movement. The temperature increase during the deformation can cause a reduction of the rolling force and rolling torque during the RCMR deformation.

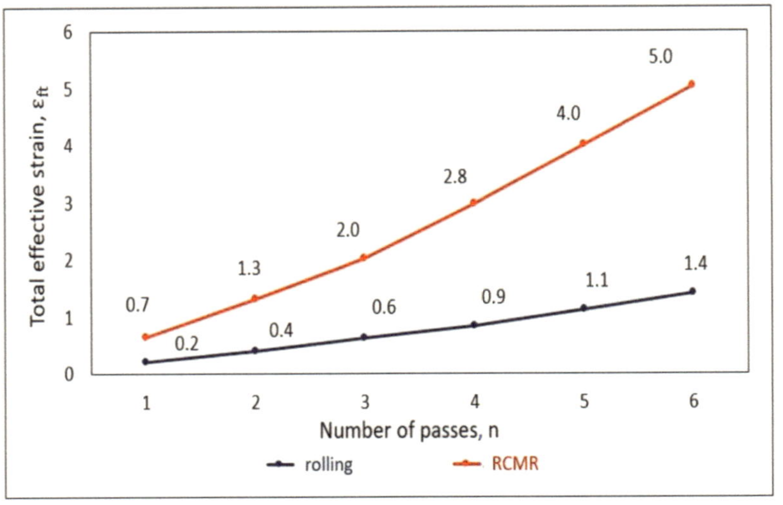

Figure 6. Changes of total effective strain in CuCr0.6 alloy (S2 state) for rolling and RCMR with number of passes. Parameters: rolling rate v = 1 (rpm), amplitude of the transverse roll movement A = 0.8 (mm), and frequency of the transverse roll movement f = 1.5 (Hz).

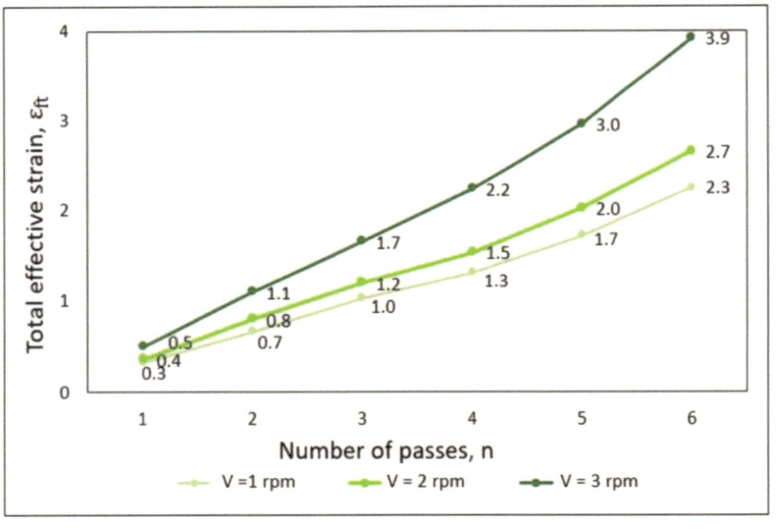

Figure 7. Changes of total effective strain in CuFe2 alloy (P state) for RCMR with number of passes and rolling rate. Parameters: amplitude of the transverse roll movement A = 0.8 (mm), and frequency of the transverse roll movement f = 1 (Hz).

The examples of the samples after RCMR processing for the rolling rates of 1, 2, and 3 rpm are presented in **Figure 9** that also shows a characteristic surface that depends on the applied parameters and could indicate the heterogeneity in the plastic deformation.

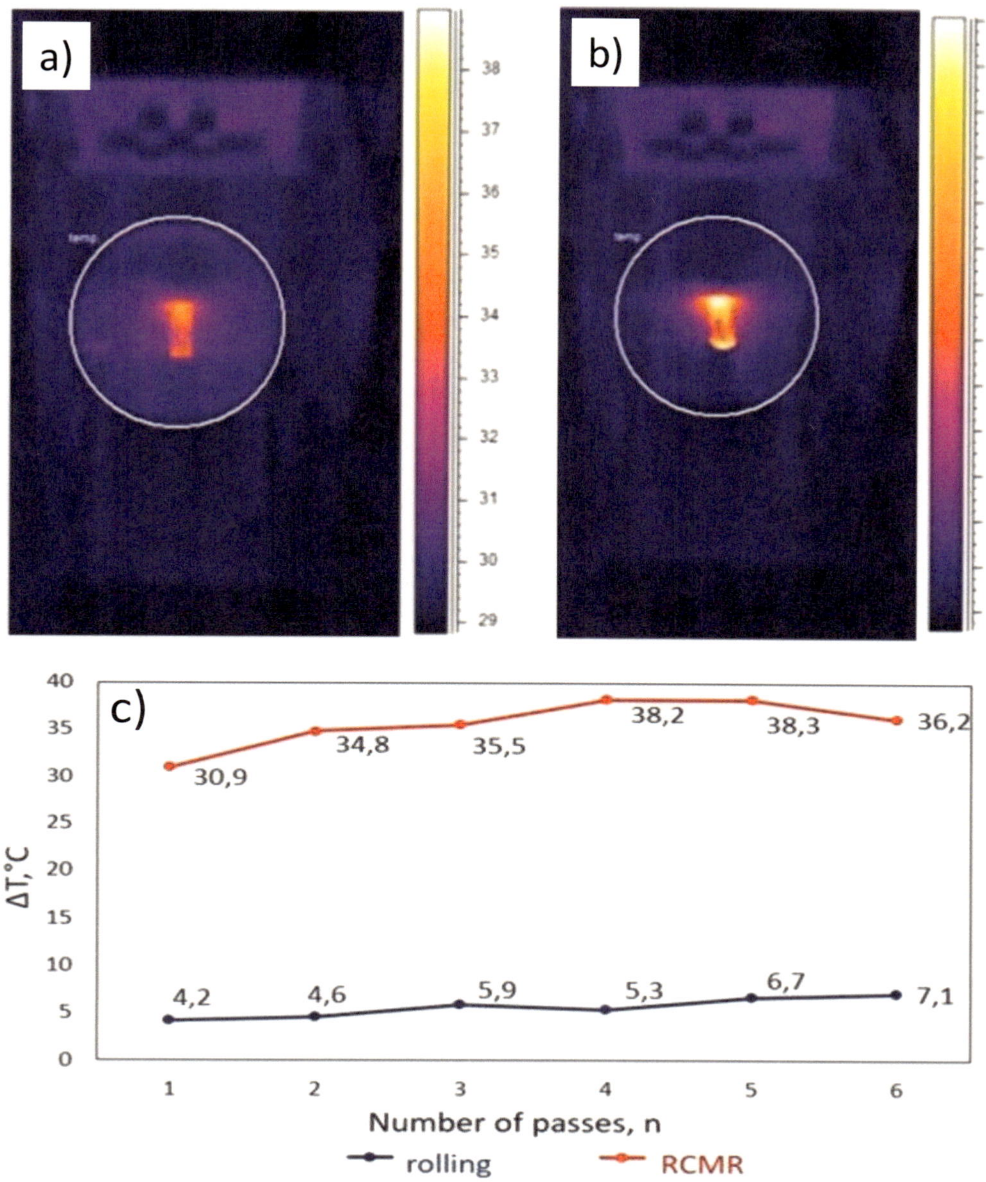

Figure 8. Thermographical images of CuFe2 (S1 state) sample surfaces after four passes obtained for: (a) rolling sample, (b) RCMR sample, and (c) changes in the temperature rise for rolling and RCMR with the number of passes. Parameters: rolling rate $v = 1$ (rpm), amplitude of the transverse roll movement $A = 0.8$ (mm), and frequency of the transverse roll movement $f = 1$ (Hz).

The transverse section of the RCMR-processed samples (**Figure 10**) exhibits inhomogeneous features along the through-thickness direction. It is clearly observed that unlike the middle, top, and bottom regions, the side region of the samples exhibits a near nondeformable structure.

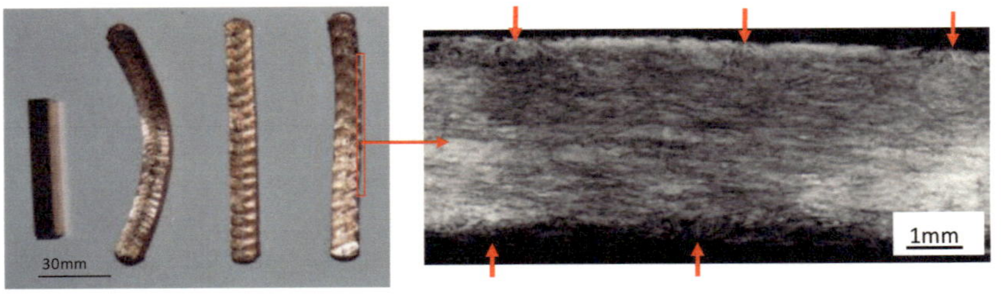

Figure 9. Macrostructure of RCMR samples before deformation and after RCMR with rolling rates of 1, 2, and 3 rpm. Characteristic "serrations" on the contact surfaces of the rollers as a result of transverse movement of rolls. Visible effects of RCMR on the lateral surface of the sample are marked by arrows.

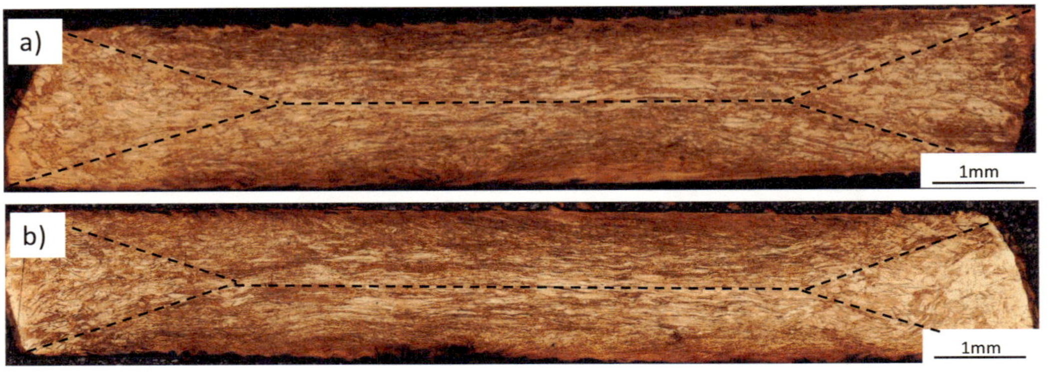

Figure 10. Transverse section of CuFe2 samples after RCMR for (a) S1 and (b) S2 states. Parameters: rolling rate $v = 1$ (rpm), amplitude of the transverse roll movement $A = 0.8$ (mm), and frequency of the transverse roll movement $f = 1.5$ (Hz), number of passes = 6.

The image in **Figure 10** also shows that the overall shape of the sample is asymmetric. This is attributed to a minor misalignment of the rolls during processing. The region of the deformed sample maintains the orientation of the as-received material (the side area), suggesting a low deformation magnitude while a different orientation of the structural features is observed in another area, suggesting a high deformation magnitude.

The heterogeneous features observed along the thickness of the samples are attributed to a difference in the deformation magnitude and can be easily seen at higher magnification (**Figure 11**).

STEM investigations confirm the heterogeneity in the refinement structure (**Figure 12**) with the microstructural observations revealing the coexistence of alternating elongated grains with a high dislocation density (**Figure 12b**)and the fine equiaxed grain without internal dislocations (**Figure 12a**).

Figure 13 shows a color-coded distribution of the hardness along the half-transverse sections of the CuFe2 alloy sample after RCMR for selected numbers of passes. It is observed that significant hardness variations are presented in the sample, with higher hardness values observed in the top and bottom areas of the sample where transverse rolling was imposed and

http://dx.doi.org/10.5772/intechopen.68954

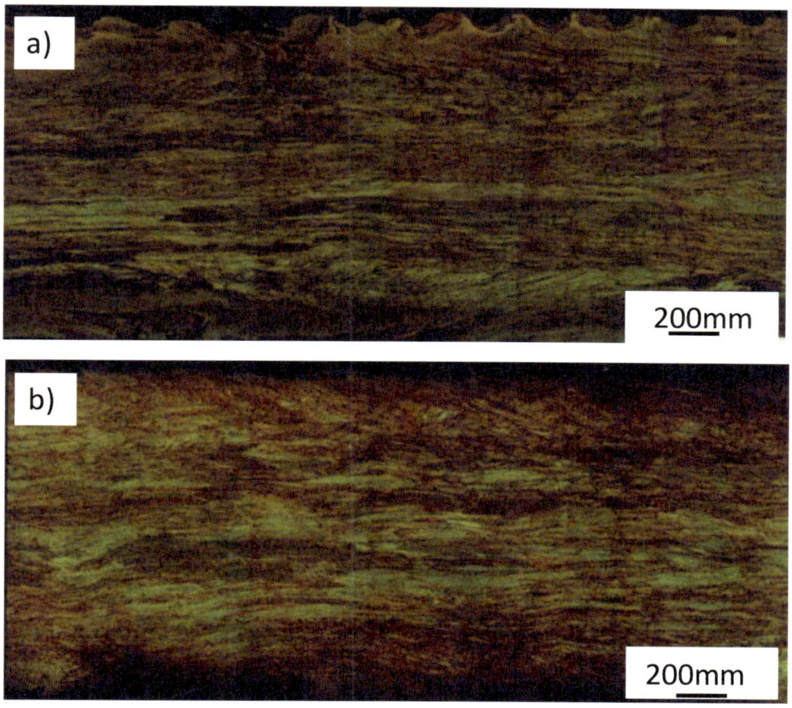

Figure 11. Transverse section of (a) CuCr0.6 sample (P state) after RCMR with parameters: rolling rate v = 0.7 (rpm), amplitude of the transverse roll movement A = 0.8 (mm), and frequency of the transverse roll movement f = 1.5 (Hz), number of passes = 6, (b) CuFe2 sample (S2 state) after RCMR with parameters: rolling rate v = 1 (rpm), amplitude of the transverse roll movement A = 0.8 (mm), and frequency of the transverse roll movement f = 1.5 (Hz), number of passes = 6.

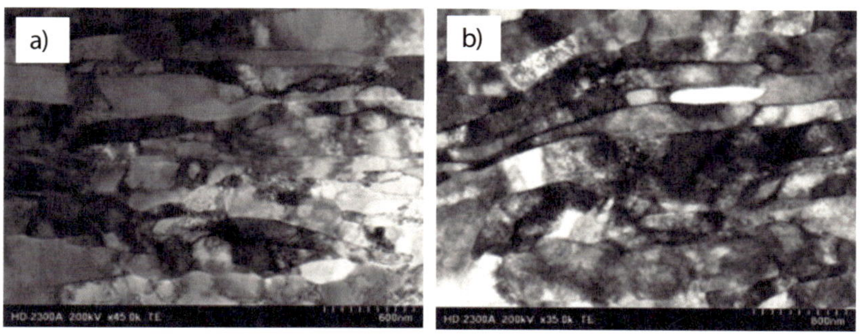

Figure 12. STEM microstructures of CuFe2 (S2 state) sample taken from different areas after RCMR with parameters: with parameters: rolling rate v = 1 (rpm), amplitude of the transverse roll movement A = 0.8 (mm), and frequency of the transverse roll movement f = 1.5 (Hz), number of passes = 6.

lower hardness values are observed in the central areas. It is important to note that the hardness distribution in the samples is in agreement with the microstructures shown in **Figures 10** and **11**.

The measured mechanical parameters for CuFe2 and CuCr0.6 alloys are presented in **Table 3** and it can be seen that solid solution treated CuFe2 and CuCr0.6 alloys that exhibit strain hardening, low strength, and high ductility. The best strength properties were obtained for the alloy in the S1 + RCMR state while the strength properties obtained for the alloys in the P + RCMR and S2 + RCMR states were comparable. The samples after the deformation show elongation to fracture (A_c) values in the 4–7% range and a uniform elongation (A_{gt}) of ~1%. These results are in accordance with the STEM observations of the microstructure and confirm that only fine coherent precipitates are active for the blocking of the dislocations. The lack of coherent precipitates in the P + RCMR and S2 + RCMR states effectively decreases the dislocation generation rate with the large particles.

In general, it should be noted that the SPD-processed materials generally have very high strength and hardness compared with conventional deformed materials. RCMR method belongs to cyclic method of deformation and from this point of view the tensile strength of materials with ultrafine grains does not become much higher compared with conventional process. Evidently, it is observed that tendency of ductility increases in materials deformed by RCMR process. This is the effect of structure formation. It is known that the plastic deformation results in microstructural refinement through formation of a three-dimensional dislocation boundary structure. The dislocation boundaries formed during conventional rolling are predominantly rotation boundaries, so that the refinement is not just spatial, but

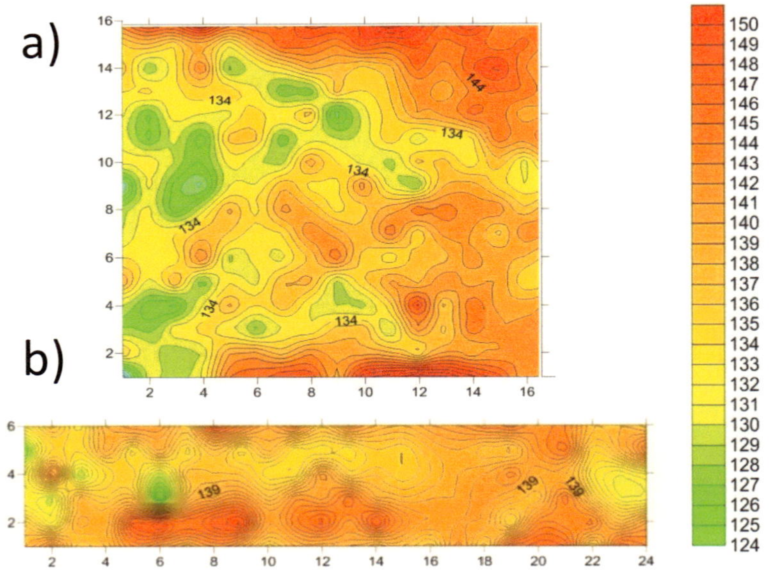

Figure 13. Distribution of measured hardness values on the transverse section of samples CuFe2 (P state) for two passes (a); and six passes (b). Parameters: rolling rate $v = 1$ (rpm), amplitude of the transverse roll movement $A = 0.8$ (mm), and frequency of the transverse roll movement $f = 1.5$ (Hz).

Material state		YS (MPa)	UTS (MPa)	A_{gt} (%)	A_c (%)	HV	IACS (%)
CuCr0.6	Initial state	97 ± 4	214 ± 5	22.5 ± 1	26.0 ± 1	43 ± 4	40
	Quenching + RCMR	363 ± 5	393 ± 9	1 ± 0.2	4.7 ± 0.5	123 ± 15	43
	Aging at 500°C/2 h + RCMR	513 ± 7	539 ± 7	1.1 ± 0.1	4.8 ± 0.6	180 ± 6.5	83
	Aging at 700°C/24 h + RCMR	324 ± 9	360 ± 5	1.2 ± 0.2	4.4 ± 0.2	132 ± 9	86
CuFe2	Initial state	148 ± 5	246 ± 4	14.2 ± 0.5	17.7 ± 0.6	60 ± 4	28
	Quenching + RCMR	356 ± 7	361 ± 6	1.0 ± 0.5	4.7 ± 0.5	126 ± 14	22
	Aging at 500°C/2 h + RCMR	388 ± 7	393 ± 6	0.9 ± 0.3	4.0 ± 0.3	150 ± 16	41
	aging at 700°C/24h+RCMR	366 ± 7	371 ± 5	1.2 ± 0.3	7.0 ± 0.3	133 ± 9	34

Table 3. Measured mechanical parameters: yield strength (YS), ultimate tensile strength (UTS), uniform elongation (A_{gt}), and elongation to fracture (A_c) of deformed CuCr0.6 and CuFe2 alloys.

also crystallographic. During RCMR deformation dislocation mobility is increased as a result of temperature increase, for example, thereby enabling the establishment of a fully three-dimensional boundary structure. The materials having such a structure are characterized by a number of specific properties including significantly higher yield point or ductility than that produced by conventional deformation methods as rolling.

The microstructure formation and the influence of the microstructure on the mechanical properties in CuFe2 alloy during the RCMR deformation have been well-documented elsewhere by the author [14]. In these studies, it has been clearly demonstrated that the grain refinement process observed on the cross-section plane is not homogeneous for all deformation states and that the impact of transverse rolling on the material is most apparent for the surface layers. It is found that the samples in all states (P, S1, S2) mostly exhibit lamellar/elongated structures with a mixture of low- and high-angle boundaries. It was also clearly shown that the deformation introduced into the material during RCMR does not guarantee the development of a refined microstructure with a high fraction of ultrafine grains with high-angle boundaries. The heterogeneity character depends on the structure in the initial state (heat treatment conditions) as schematically shown in **Figure 14**. The differences in the grain refinement can be understood by considering the intensity of the RCMR deformation of the sample in the cross section. For the S1 sample, the microstructural refinement is less pronounced in the volume of the sample, because coherent particles strongly affect the grain/subgrain size refinement. The lack of efficient barriers in the form of a high amount of dislocations leads to the comparable intensities of RCMR deformation for the cross sections of the P and S2 samples.

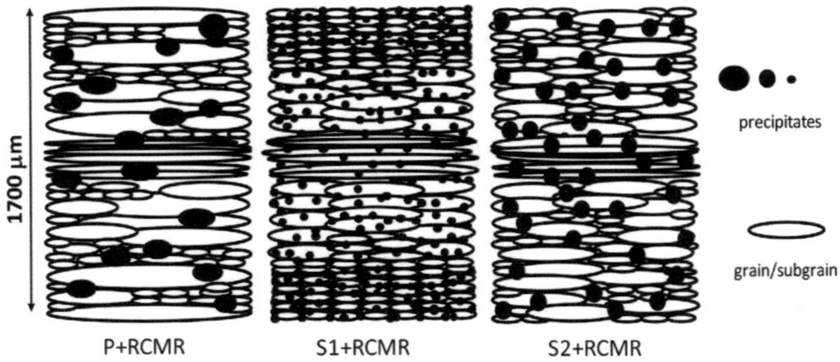

Figure 14. Scheme of microstructure refinement in RCMR processed samples.

5. Production of ultrafine grained structure of CuCr0.6 and CuFe2 alloys by COT method

Compression with oscillatory torsion (COT) is the second method patented by Silesian University of Technology, Faculty of Materials Engineering and Metallurgy in Poland and discussed in this paper. This processing technique can obtain a sub-micrometer grain size in a short time of deformation. **Figure 15a** and **b** present, respectively, a schematic of the COT setup as well as the illustration of the deformation of the samples by the simultaneous application of compression and oscillatory torsion which is the basic principle of the COT method. The method is characterized by the heterogeneity of deformation with the most intense deformations occurring at locations that are closest to the lateral surfaces of the material resulting from the application of the torsional moment. The oscillating course of the flow stress σ_p in the range σ_{pmin}–σ_{pmax} is caused by the cyclical variation of the torque M_t from 0 to M_{tmax} (**Figure 15c**).

These SPD techniques were adopted for grain size refinement in Cu and Al [15, 17].

Due to the complexity of the deformation process, the effective strains were calculated following the Huber-Mises-Hencky method and are given by

$$\varepsilon_f = \sqrt{\varepsilon^2 + \frac{\gamma^2}{3}}, \tag{4}$$

where ε is the deformation induced by the uniaxial strain and γ is the shearing strain.

The total effective strain ε_{ft} is expressed as the sum of effective strains obtained from Eq. (4) in a single phase of deformation as:

$$\varepsilon_{ft} = \sum_{n=1}^{n} \varepsilon_{fn}, \tag{5}$$

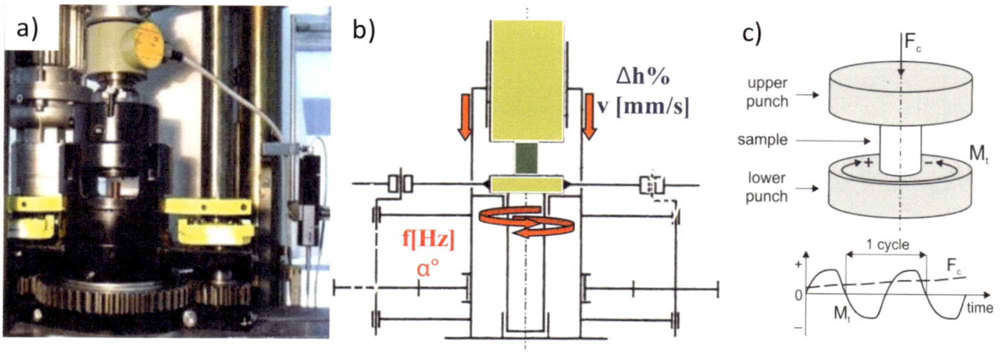

Figure 15. (a) Setup of COT, (b) schematic illustration of COT, and (c) scheme of the sample.

where n is the number of the deformation phases. The single phase comprises a torsion of the sample in one direction with a simultaneous decrease of the height.

The value of the total effective strain ε_{ft} can be controlled by changing the proportions of the following parameters: torsion frequency f in the range of 0–1.8 (Hz), compression rate v maximal 0.66 (mm/s), torsion angle α in the range of 0 to ±8 (°), and true reduction Δh (mm). **Table 4** shows the dependence of the changes in the compression force on the initial state of alloys for COT deformation. For all samples, the values of the compression force are larger for S1 state.

To illustrate the heterogeneity in the mechanical properties resulting from the COT deformation for the samples with different initial states (heat treatment), hardness distribution maps were created for selected samples. **Figures 16** and **17** show a color-coded hardness distribution along the half-longitudinal sections of the CuCr0.6 alloy sample prior to the deformation and after COT processing at ε_{ft} = 38, respectively. It is observed that the hardness variation depends on the distance from the center (**Figure 17c–e**). Higher hardness levels are observed around the 0–0.5r areas where influence of the compression process can be seen clearly. Lower hardness magnitudes are observed near the 0.5–1r distance from the center.

The higher hardness at a moderate distance from the center is typical for elongated structures with a high dislocation density. This region is quite different from the region near the surface with the refined equiaxed grains/subgrains. With the increase of the deformation ε_{ft} = 61, the homogenization of the hardness variation is not observed (**Figure 18**). Examination of the hardness distribution shows that local microareas with smaller hardness values are present in the processed samples. Therefore, the samples produced with COT exhibit heterogeneous hardness along the thickness of the samples. The increased deformation value has a moderate influence on the hardness distribution along the thickness of the sample. The increase of the deformation from ε_{ft} = 38 to 61 is not very effective for increasing the hardness. The Vickers hardness corresponds rather well to the microstructure evolution and the obtained results suggest that the Vickers hardness is affected by the dislocation structure rather than by the grain/subgrain sizes and boundaries misorientations. A coarse structure with a highly developed dislocation density gives hardness values that are larger than those for a well-defined fine-grained structure with a lower dislocation density.

Heat treatment	**CuFe2**		**CuCr0.6**	
	ε_f **= 38**	ε_f **= 61**	ε_f **= 38**	ε_f **= 61**
	Compression force *F* [kN]			
P + COT	22.7	58.0	14.3	46.9
S1 + COT	33.0	60.5	39.7	61.4
S2 + COT	24.8	58.0	25.0	58.3

Table 4. Dependence of the changes in the compression force on the initial state of alloys.

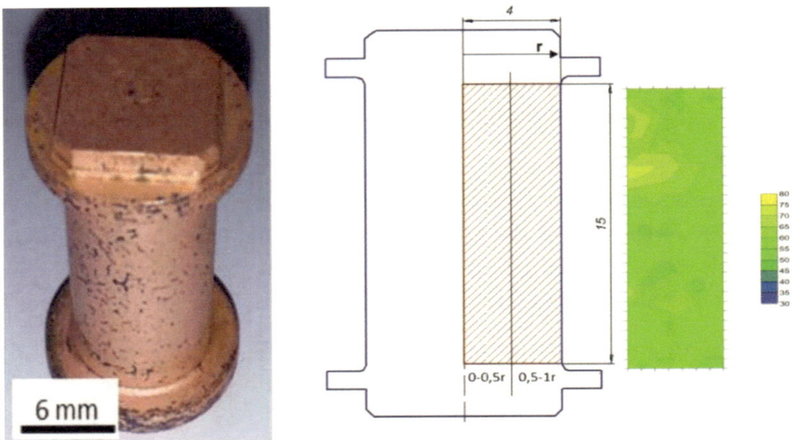

Figure 16. Distribution of measured hardness values on the longitudinal section of CuCr0.6 sample in initial (P) state before deformation.

Figure 19 shows the example of EBSD maps of the CuCr0.6 samples deformed at ε_{ft} = 45. Even though the original grains were plastically deformed, the grains are not significantly elongated perpendicular to the compression direction. This means that the compression does not have a strong effect on the deformed microstructure. The original grain boundaries are very jagged (the grains are comprised of a partially bulged original grain boundary) and the deformed grains are delineated by irregular bands due to the deformation occurring on different slip systems. Intersections of the deformation bands give rise to elongated grains/subgrains with mostly low-angle boundaries with misorientation angles largely in the 2–5° range, indicating a high density of cell and subgrain structures (**Figure 19a** and **b**).

In other areas, the refinement microstructure became better organized and the new boundaries were characterized by misorientation angles in the 5–15° range (**Figure 20a**). These microstructures involve the annihilation and rearrangement of the dislocations and were characterized for the P and S2 states (**Figures 19a**, **b**, **e**, **f**, and **20c**). **Figure 19c** and **d** reveals that the density of low-angle boundaries was significantly reduced and grains with boundary angles higher than 15° were found in the microstructure. The microstructure in state S1

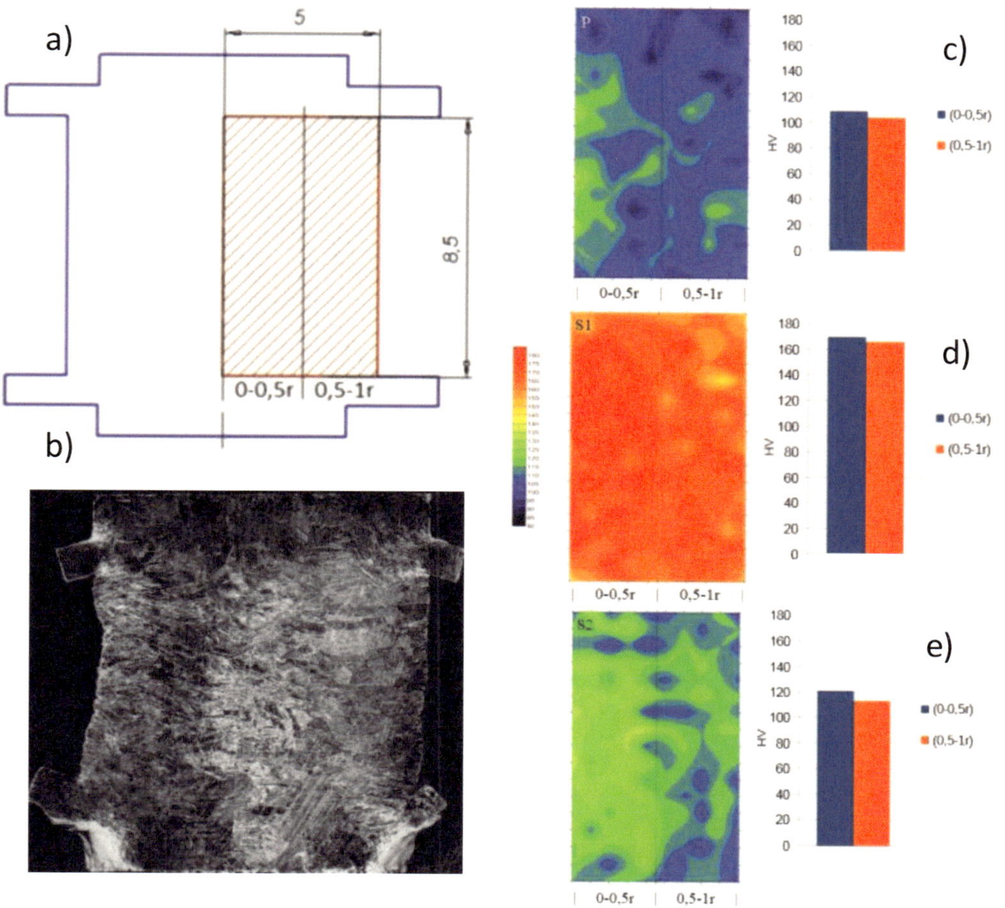

Figure 17. Distribution of measured hardness values on the longitudinal section (a, b) of CuCr0.6 samples after COT with parameters: torsion frequency $f = 0.8$ (Hz), compression rate $v = 0.015$ (mm/s), torsion angle$\alpha = \pm 6$ (°), and height reduction $h = 50$ (%). Total effective strain ($\varepsilon_f = 38$) for different initial structures; (c) P state; (d) S1 state; (e) S2.

is quite heterogeneous because a number of subgrains with low misorientation angles were observed in addition to the ultrafine grains with high misorientation angles (**Figure 20b**).

COT deformation with $\varepsilon_{ft} = 12$ typically exhibits the problem to eliminate the low-angle boundaries and transform the microstructure into one with higher angle boundaries (**Figure 21**).

The obtained results were confirmed by STEM investigations for $\varepsilon_{ft} = 28$and 45 (**Figure 22**). The COT process produces well-defined subgrain structures with low- and medium-angle misorientations that were identified based on the diffraction contrast (**Figure 22a**). With increasing strain, the structures observed in **Figure 22a** were comparable but additionally, a recovery process was observed in some microareas as shown in **Figure 22b**. This effect may be connected to the specifics of the deformation realized by the cyclic deformation. It is also possible

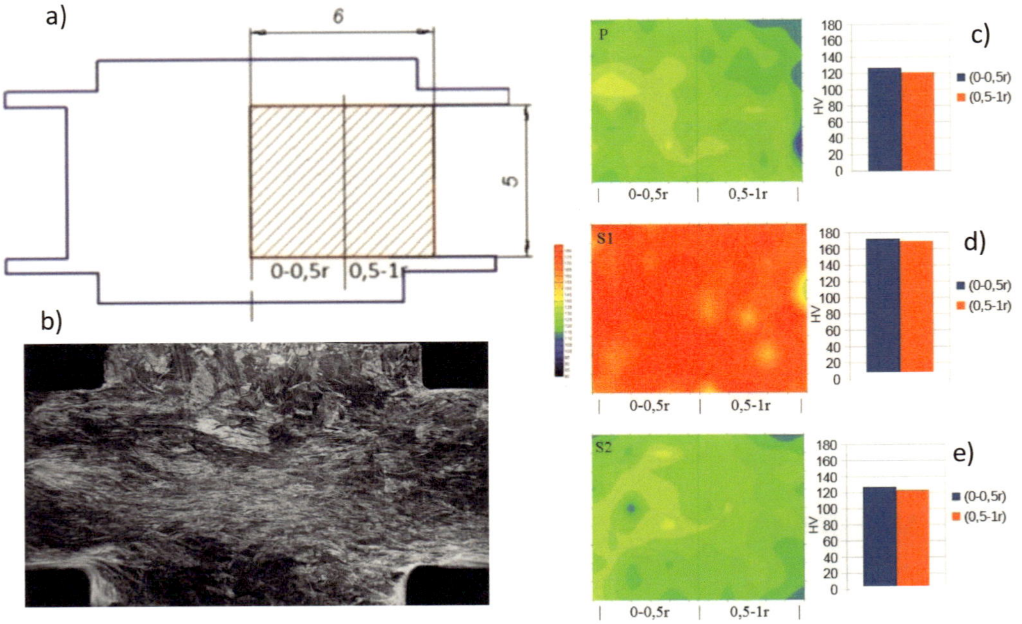

Figure 18. Distribution of measured hardness values on the longitudinal section (a, b) of CuCr0.6 samples after COT with parameters: torsion frequency $f = 0.8$ (Hz), compression rate $v = 0.015$ (mm/s), torsion angle $\alpha = \pm 6$ (°), and height reduction $h = 80$ (%). Total effective strain ($\varepsilon_f = 61$) for different initial structures (c); P state; (d) S1 state; (e) S2 state.

that the temperature increased during the deformation and thus only an insignificant effect on the grain refinement is observed. The absence of small precipitates results in the effective loss of dislocations, which is unfavorable during the COT deformation. The lack of effective barriers for dislocation pinning gives rise to intensive recovery processes (**Figures 19a**, **b** and **22a**, **b**). For sample S1, the microshear banding contributes to the grain subdivision and the subgrain/grain structure dominates within the deformed bands (**Figure 22c**). Additionally, tangled dislocations are observed in the matrix. **Figure 22d** shows the microstructure of the CuCr0.6 alloy in the S1 state after the deformation at $\varepsilon_{ft} = 45$; it can be seen that the microstructure becomes more distinct and difficult to resolve. Many of the boundaries are not well-defined and are curved instead of straight, indicating the presence of a high internal stress. This type of microstructure is attributed to the development of arrays of high-energy non-equilibrium boundaries. Many dislocations are visible at both the grain boundaries and inside the grains. The typical microstructure for the S1 state becomes more heterogeneous than that of the P state (compare **Figures 22c**, **d** and **a**, **b**). With the increase of the deformation, the heterogeneity of the structure in the sample is still visible. This means that in this state, it is still more difficult for the high-energy boundaries to transform into stable arrays. We observe a very fine microstructure composed of grains that are smaller than 200 nm in sample S1. The microstructures of samples S2 processed for $\varepsilon_{ft} = 28$and $\varepsilon_{ft} = 45$ are shown in **Figures 22e** and **f**, respectively. The larger grains/subgrains coexist with small grains/subgrains. There are almost no dislocations in some grains/subgrains, and chaotically distributed dislocations

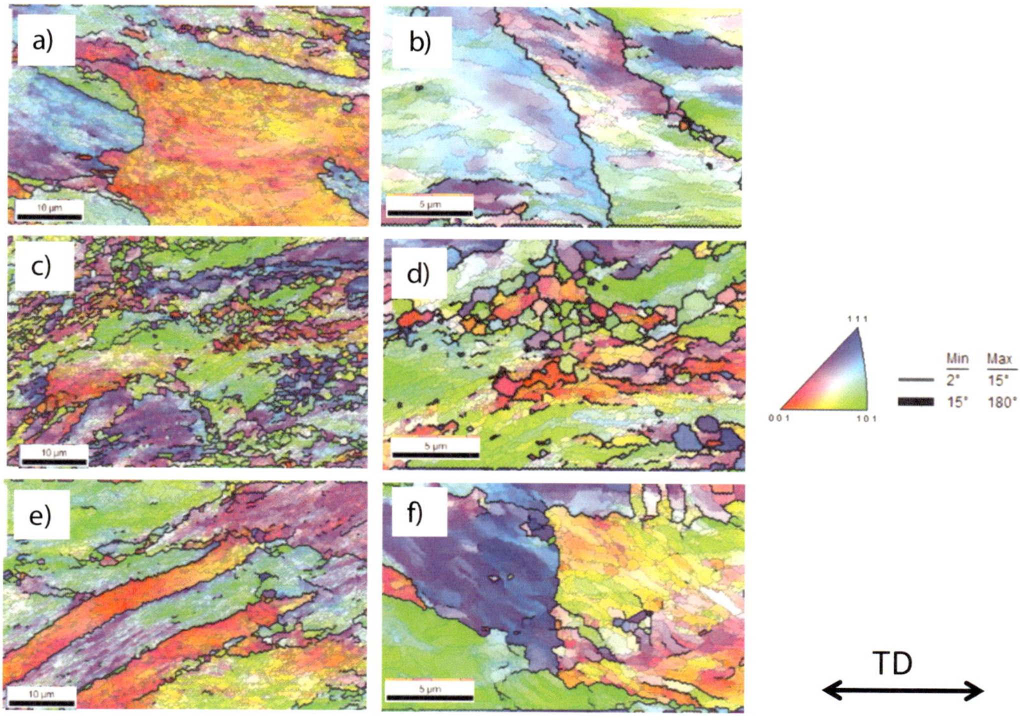

Figure 19. Orientation imaging maps ofCuCr0.6 sample showing deformed microstructures under different conditions: (a, b) P + COT state; (c, d) S1 + COT state; and (e, f) S2 + COT state. Deformation parameters: torsion frequency $f = 1.6$ (Hz), compression rate $v = 0.04$ (mm/s), torsion angle $\alpha = \pm 6$ (°), and height reduction $h = 80$ (%). Total effective strain (ε_f = 45).

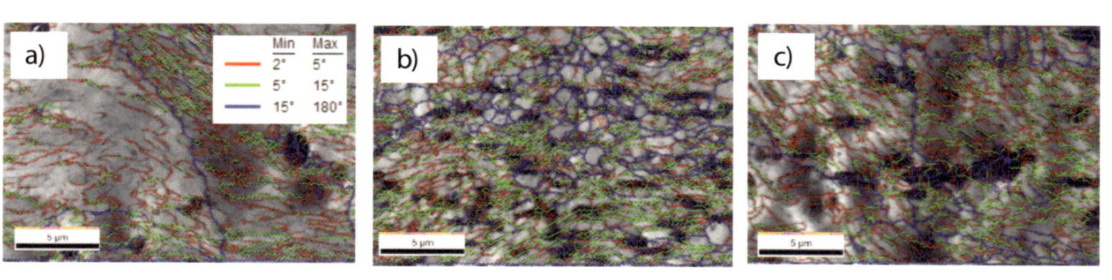

Figure 20. EBSD maps with different types of grain boundaries in CuCr alloys (a) P state, (b) S1 state, (c) S2 state. Deformation parameters: torsion frequency $f = 1.6$ (Hz), compression rate $v = 0.04$ (mm/s), torsion angle $\alpha = \pm 6$ (°), and height reduction $h = 80$ (%). Total effective strain (ε_f = 45).

are present in other large or small grains/subgrains. The average grain size was refined to about ~300–500 nm. From the literature [3–7], it is known that the microstructural evolution depends mainly on the strain magnitude, with a more homogeneous microstructure obtained with increasing strain. In our results, the differences in the microstructures are still observed even though the magnitude of the deformation increases. In general, it should be noted that it is impossible to obtain a homogeneous structure using this method.

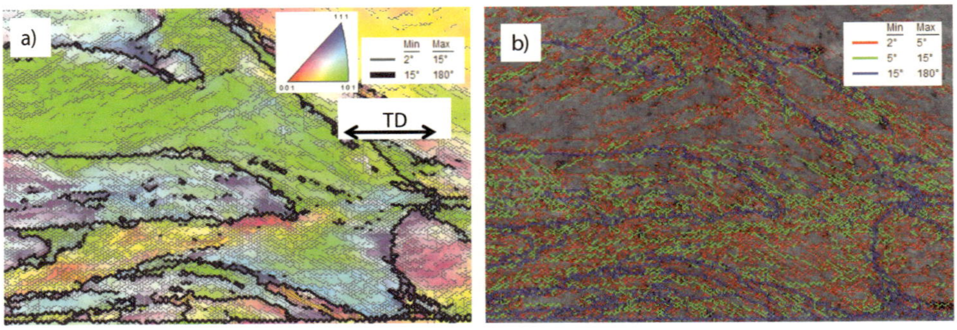

Figure 21. Orientation imaging map (a); and EBSD maps with different types of grain boundaries (b) for CuFe2 alloy after COT deformation with parameters: torsion frequency f = 1.6 (Hz), compression rate v = 0.1(mm/s), torsion angle α = ±6 (°), and height reduction h = 80 (%). Total effective strain (ε_f = 12).

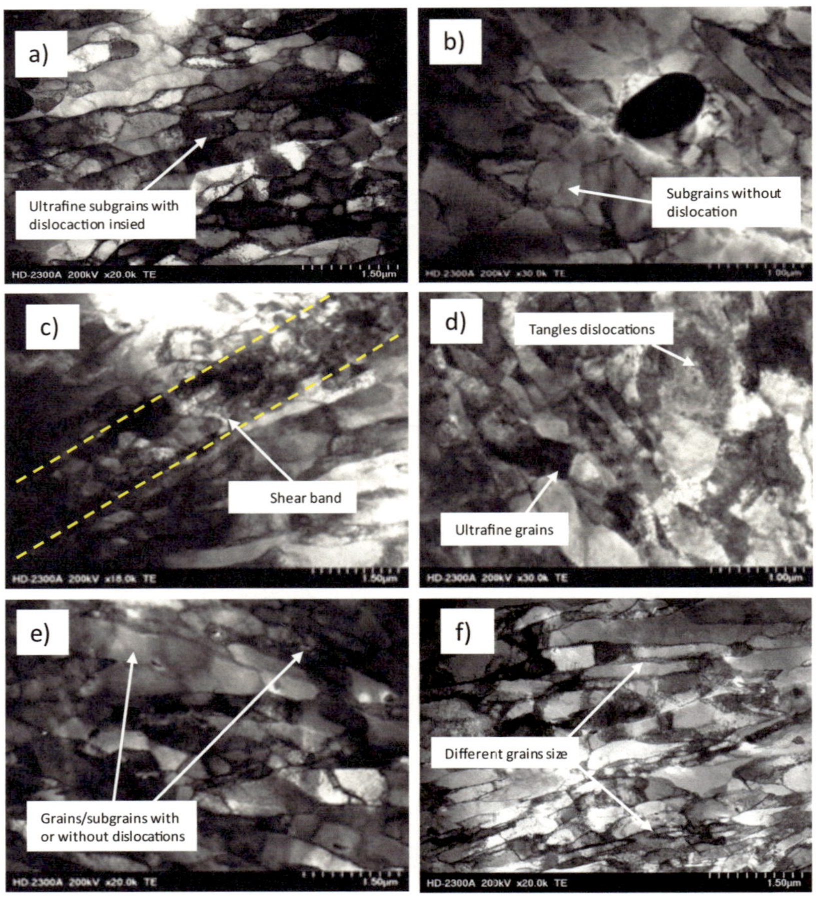

Figure_22.

Figure 22. STEM microstructures for CuCr0.6 alloy after COT deformation with parameters: (a) P state, ε_{ft} = 28, (b) P state, ε_{ft} = 45, (c) S1 state, ε_{ft} = 28, (d) S1 state, ε_{ft} = 45, (e) S1 state, ε_{ft} = 28, (f) S1 state, ε_{ft} = 45.

For the S1 state, COT processing hinders the deformation processes for both CuCr0.6 and CuFe2 alloys. For example, this can be clearly seen for $\varepsilon_{ft} = 28$ where coherent precipitates reduce the number of operating slip systems (**Figure 23a** and **b**). The much more equiaxed structures (S2 state) visible in **Figure 23c** and **d** are due to the occurrence of different slip systems.

The example distribution of the grain/subgrain diameter values for the CuCr0.6 alloy (P and S1 state) for $\varepsilon_{ft} = 45$ is shown in **Figure 24**.

It is important to note that based on SEM and STEM observations, the grain boundaries obtained during COT deformation are in the nonequilibrium state. It has been argued that such boundaries provide a large number of excess dislocations for the slip systems and can enable the grains to rotate at room temperature, leading to a significant increase in the strain hardening and ductility. The results of mechanical properties for CuCr0.6 and CuFe2 alloys are shown in **Figure 25** and in **Table 5**.

Several articles recently reported UFG materials maintaining both a high strength and an adequate ductility [6–9]. An especially high strength and good ductility in ultrafine-grained materials produced by the COT deformation were obtained for the S1 state. The fine precipitates interacted with the dislocations and exhibited a strong pinning effect on the dislocation movement, leading

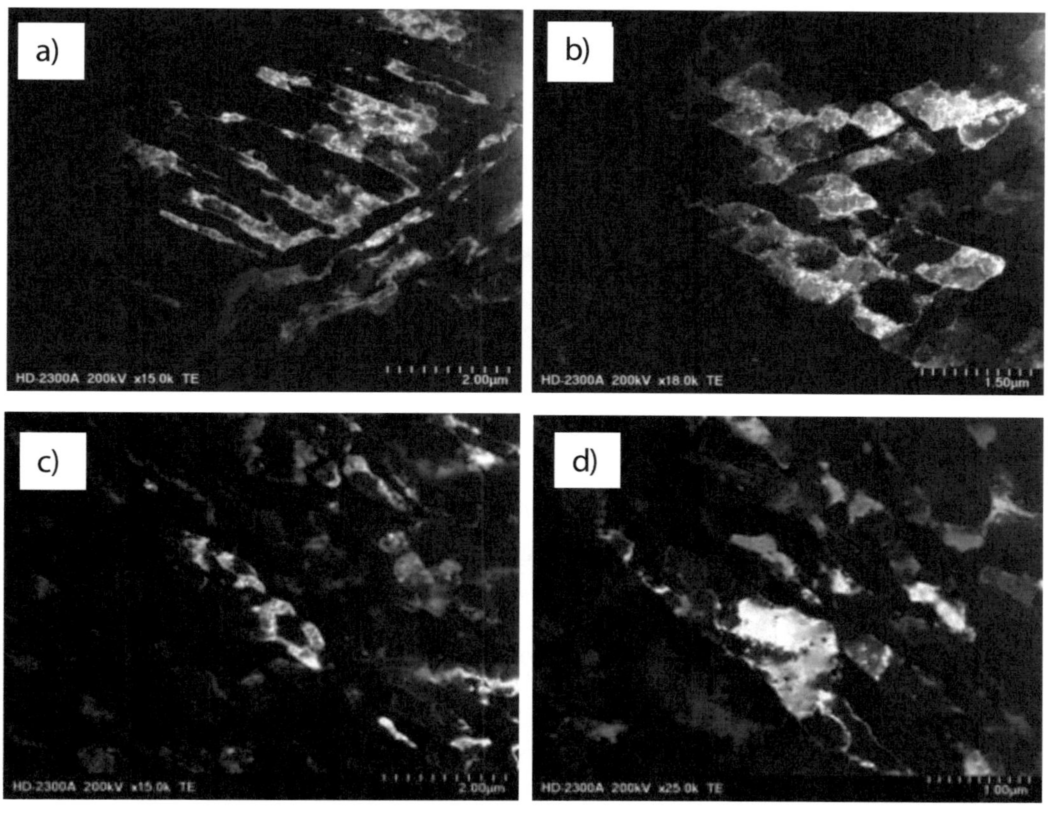

Figure 23. Dark field images obtained from STEM investigations for CuFe2 alloys after COT deformation with $\varepsilon_{ft} = 28$: (a, b) S1 state, (c, d) S2 state.

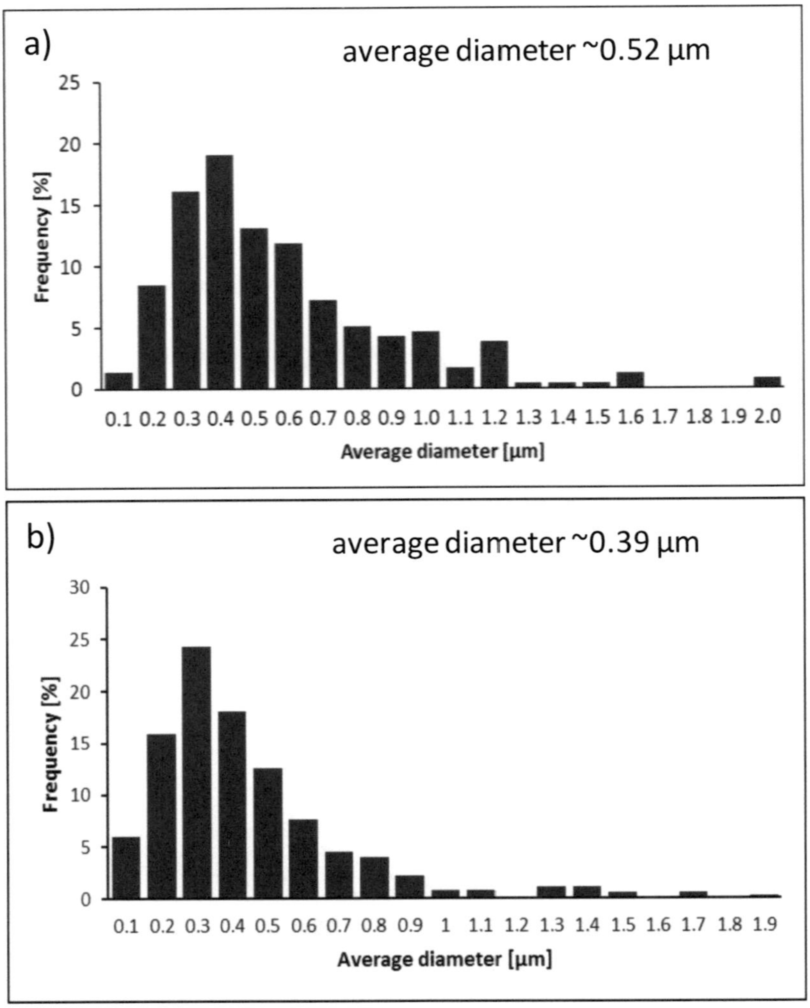

Figure 24. Grain/subgrain size distribution for CuCr0.6 alloy after COT deformation for $\varepsilon_{ft} = 45$: (a) P state, (b) S1 state.

to high yield strength (YS) and ultimate tensile strength (UTS) values. For the S2 state, the precipitates coarsened remarkably to an average size of 100 nm with a larger interparticle spacing. The coalescence of the precipitates accompanied by an enhanced recovery process resulted in a significant flow stress decrease. It is important to note that SPD processing leads to a reduction in the ductility that is generally smaller than those for the more conventional deformation processing techniques such as rolling, drawing, and extrusion. The low ductility is caused by the low strain-hardening rate, giving rise to early localized deformation in the form of necking. The availability of a bimodal (a mixture of two or multiphases with varying scales and properties) grain size distribution leads to a considerable increase in the ductility. We note that based on the obtained results, a significant degradation in the ductility of the specimens is typical for COT. The YS and UTS values are for the two alloys, comparable when are taken into account the heat

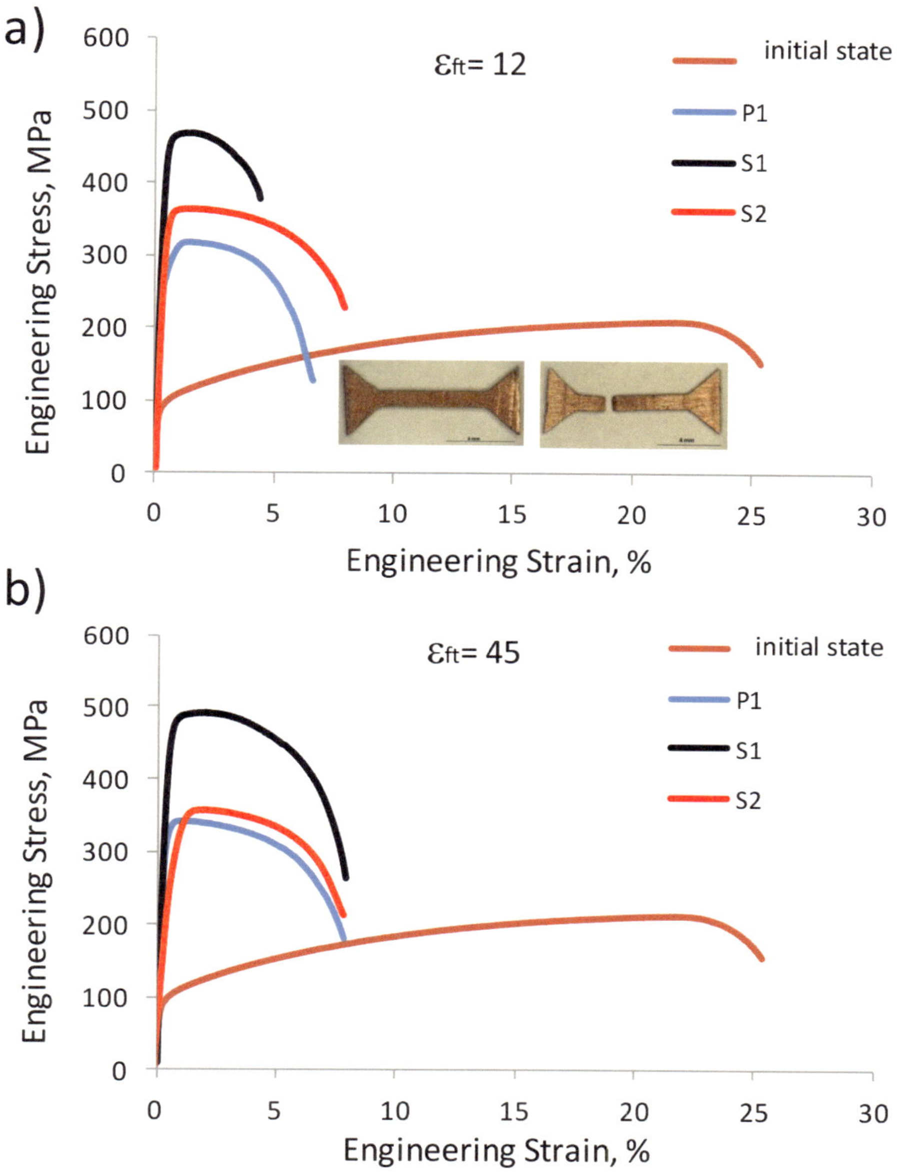

Figure 25. Engineering strain versus engineering stress for CuCr0.6 alloys after application of different magnitudes of total effective strain. a) $e_{ft} = 12$ and b) $e_{ft} = 45$.

treatment processes (maximal value of YS and UTS for S1 state), implying that a high content of Cr and Fe solute atoms in the Cu matrix of the deformed sample reduces the dislocation mobility and retards the dynamic recovery at a level comparable to that of the noncoherent but small particles observed for the S2 state. The pinning of the dislocation by these precipitates is much weaker but it does effectively influence the pinning of the grain/subgrain boundaries.

Material state		YS (MPa)	UTS (MPa)	A_{gt} (%)	A_c (%)	HV	IACS (%)
CuCr0.6	Initial state	97 ± 4	214 ± 5	22.5 ± 1	26.0 ± 1	43 ± 4	40
	Quenching + COT	336 ± 6	342 ± 6	1 ± 0.2	8 ± 0.8	123 ± 15	40
	Aging at 500°C/2 h + COT	464 ± 11	491 ± 3	1.8 ± 0.1	4.3 ± 0.3	180 ± 6.5	85
	Aging at 700°C/24 h + COT	343 ± 8	360 ± 7	1.7 ± 0.5	8.8 ± 1.5	132 ± 9	86
CuFe2	Initial state	148 ± 5	246 ± 4	14.2 ± 0.5	17.7 ± 0.6	60 ± 5	28
	Quenching + RCMR	329 ± 31	340 ± 34	1.8 ± 0.4	7.8 ± 1.6	126 ± 17	26
	Aging at 500°C/2 h + COT	378 ± 18	387 ± 17	1.6 ± 0.3	8 ± 1.4	150 ± 15	48
	Aging at 700°C/24 h + COT	309 ± 10	331 ± 17	2.6 ± 0.8	9.5 ± 1.8	133 ± 9	34

Table 5. Measured mechanical parameters: yield strength (YS), ultimate tensile strength (UTS), uniform elongation (A_{gt}), and elongation to fracture (A_c) of deformed CuCr0.6 and CuFe2 alloys using COT for $\varepsilon_{ft} = 45$.

6. Summary

The results obtained for the CuFe and CuCr alloys deformed through RCMR and COT indicated the possibility of obtaining materials with the UFG structure and high mechanical properties. During the RCMR and COT deformations, the microstructure was significantly refined but was heterogeneous after the application of a high total effective strain. Initially, the first process of deformation (lower value of total effective strain) dominates the development of deformation-induced boundaries that occur heterogeneously in the volume according to the net strain gradient. The measured grain/subgrain sizes are in the 200–500 nm range, with a mixture of low- and high-angle boundaries. In contrast to the commonly found statements in the literature, subsequent deformations (increased total effective strain) do not induce an increase of the grain misorientation and a decrease in the grain/subgrain size. Continuous heterogeneity in the microstructure increases in both alloys as a result of the cyclical character of these processes. Cyclic character of these methods promotes dynamic recovery, involving the decrease of dislocation density in the interior of grains/subgrains and in grain/subgrain boundaries.

The initial structure (heat treatment process) of the alloy influences the RCMR and COT deformations. The best combination of mechanical properties was obtained for the material deformed in the S1 state. The Fe and Cr precipitates in the S1 state that is responsible for the peak-aged strength of CuFe and CuCr alloys are expected to pin the dislocations. While the noncoherent particles in the S2 state are still capable of effectively pinning the grain boundaries and prevent further coarsening, they are not as effective as the precipitates in the S1 state. The deformation sample in the P state shows hardening that is comparable to that of the S2 state and the deformation of the P samples results in a structure recovery.

The present investigation confirms that the cyclic deformation (RCMR and COT method) plays a significant role in the structure evolution. Especially, these processes decrease the rate of high angle boundaries (HABs) formation at the stages of large strain accumulation. Moreover, RCMR and COT deformations increase in the structure in homogeneity with increasing strain. Additionally, cyclic deformation processes did not affect, significantly, the effective grain refinement with increasing strain.

Acknowledgements

This work was supported by the National Science Centre in Poland under contract No. UMO-2013/09/B/ST8/01695.

Author details

Kinga Rodak

Address all correspondence to: kinga.rodak@polsl.pl

Faculty of Materials Engineering and Metallurgy, Silesian University of Technology, Katowice, Poland

References

[1] Prangnell PB, Bowen JR, Apps PJ. Ultra-fine grain structures in aluminium alloys by severe deformation processing. Materials Science and Engineering: A. 2004;**375-377**:178-185. DOI:10.1016/j.msea.2003.10.170

[2] Shaarbaf M, Toroghinejad MR. Nano-grained copper strip produced by accumulative roll bonding process. Materials Science and Engineering: A. 2008;**473**:28-33. DOI: 10.1016/j.msea.2007.03.065

[3] Chinh NQ, Horvath G, Horita Z, Langdon TG. A new constitutive relationship for the homogeneous deformation of metals over a wide range of strain. Acta Materialia. 2004;**52**:3555-3563. DOI: 10.1016/j.actamat.2004.04.009

[4] Aldazabal J, Sevillano JG. Hall-Pech behaviour induced by plastic strain gradiens. Materials Science and Engineering: A. 2004;**365**:186-190. DOI: 10.1016/j.msea.2003.09.026

[5] Mishra A, Kad BK, Gregori F, Meyers MA. Microstructural evolution in copper subjected to severe plastic deformation: Experiments and analysis. Acta Materialia. 2007;**55**:13-28. DOI: 10.1016/j.actamat.2006.07.008

[6] Orlov D, Beygelzimer Y, Synkov S, Varyukhin V, Tsuji N, Horita Z. Microstructure evolution in pure Al processed with twist extrusion. Materials Transactions. 2009;**50**:96-100. DOI: 10.2320/matertrans.MD200802

[7] Huang WH, Yu CY, Kao PW, Chang CP. The effect of strain path and temperature on the microstructure developed in copper processed by ECAE. Materials Science and Engineering: A. 2004;**366**:221-228. DOI: 10.1016/j.msea.2003.08.033

[8] Dobatkin SV, Szpunar JA, Zhilyaev AP, Cho JY, Kuznetsov AA. Effect of the route and strain of equal-channel angulat pressing on structure and properties of oxygen-free copper. Materials Science and Engineering: A. 2007;**462**:132-138. DOI: 10.1016/j.msea.2006.04.156

[9] Sun PL, Yu CY, Kao PW, Chang CP. Microstructural characteristics of ultrafine-grained aluminium produced by equal channel angular extrusion. Scripta Materialia. 2002;**47**:377-381. PII: S1359-6462(02)00117-3

[10] Ferry M, Burhan N. Structural and kinetic aspect of continuous grain coarsening in a fine-grained Al-0.3Sc alloy. Acta Materialia. 2007;**55**:3479-3491. DOI: 10.1016/j.actamat.2007.01.047

[11] Apps PJ, Berta M, Prangnell PB. The Effect of dispersoids on the grain refinement mechanisms during deformation of aluminium alloys to ultra-high strains. Acta Materialia. 2005;**53**:499-511. DOI: 10.016/j.actamat.2004.09.042

[12] Valipour Sh, Eivani AR, Jafarian HR, Seyedein SH, Aboutalebi MR. Effect of pre-deformation termomechanical processing on the development of ultrafine grain structure during equal channel angular extrusion. Materials & Design. 2016;**89**:377-384. DOI: 10.1016/j.matdes.2015.09.161

[13] Quadir MZ, Ferry M, Al-Buhamad O, Munroe PR. Shear banding and recrystallization texture development in a multilayered Al alloy sheet produced by accumulative roll bonding. Materials & Design. 2009;**57**:29-40. DOI: 10.1016/j.actamat.2008.08.056

[14] Rodak K, Radwański K. Influence of precipitates on the grain refinement in CuFe2 alloy processed by rolling with cyclic movement of rolls. Materials and Design. 2016;**110**:255-265. DOI: 10.1016/j.matdes.2016.07.138

[15] Rodak K, Pawlicki J. Microstructure characterization of Cu processed by compression with oscillatory torsion. Materials Characterization. 2014;**94**:37-45. DOI. 10.1016/j.matchar.2014.05.002

[16] Rodak K, Pawlicki J.Effect of compression with oscillatory torsion processing on structure and properties of Cu. Journal of Materials Science & Technology. 2011;**27**:1083-1088. http://www.jmst.org/EN/Y2011/V27/I11/1083

[17] Rodak K, Pawlicki J. Efficiency of the compression with oscillatory torsion method in grain refinement in Al. Archives of Civil and Mechanical Engineering. 2016;**16**:805-812. DOI. 10.1016/j.acme.2016.05.004

[18] Rodak K, Radwański K, Molak R, Pakiela Z. Studies of Cu after plastic deformation. Kovove Materialy Metalic Materials. 2010;**48**:313-319. DOI: 10.4149/km_2010_5_313

[19] Rodak K, Molak RM, Pakieła Z. Structure and properties of copper after large strain deformation. Physica Status Solidi (C). 2010;**5**:1351-1354. DOI: 10.1002/pssc.200983392

[20] Dobatkin SV, Gubicza J, Shangina DV, Bochvar NR, Tabachkova NY. High strength and good electrical conductivity in Cu-Cr alloys processed by severe plastic deformation. Materials Letters. 2015;**153**:5-9. DOI: 10.1016/j.matlet.2015.03.144

[21] Vinogradov A, Ishida T, Kitagawa K, Kopylov VI. Effect of strain path on structure and mechanical behaviour of ultra-fine grain Cu-Cr alloy produced by equal-channel angular pressing. Acta Materialia. 2005;**53**:2181-2192. DOI:10.1016/j.actamat.2005.01.046

[22] Islamgaliev RK, Nesterov KM, Bourgon J, Champion Y, Valiev RZ. Nanostructured Cu-Cr alloys with high strength and electrical conductivity. Journal of Applied Physics. 2014;**115**:194301-194304. DOI: 10.1063/1.4874655

[23] Wei KX, Wei W, Wang F, Du QB, Alexandrov IV, Hu J. Microstructure, mechanical properties and electrical conductivity of industrial Cu-0.5%Cr alloy processed by severe platic deformation. Materials Science and Engineering: A. 2011;**528**:1478-1484. DOI:10.1016/j.msea.2010.10.059

[24] Cao H, Min JY, Wu SD, Xian AP, Shang JK. Pinning of grain boundaries by second phase particles in equal-channel angularly pressed Cu_Fe_P alloy. Materials Science and Engineering: A. 2006;**431**:86-91. DOI: 10.1016/j.msea.2006.05.081

Chapter 6

Thermal Stability of Ultra-Fine Grained Microstructure in Mg and Ti Alloys

Jitka Stráská, Pavel Zháňal, Kristína Václavová,
Josef Stráský, Petr Harcuba, Jakub Čížek and
Miloš Janeček

Additional information is available at the end of the chapter

http://dx.doi.org/10.5772/intechopen.68956

Abstract

This chapter reviews the thermal stability of ultra-fine grained (UFG) microstructure in selected magnesium and titanium-based materials prepared by severe plastic deformation (SPD). The focus is on the wide palette of experimental methods applicable for investigation of microstructural stability. These methods include scanning electron microscopy (SEM), electron backscatter diffraction (EBSD), microhardness measurement, positron annihilation spectroscopy (PAS), and electrical resistance measurement. Microstructural stability of UFG commercially pure (CP) Ti and Ti-6Al-7Nb alloy produced by equal-channel angular pressing (ECAP) is studied *ex situ* after annealing by SEM, by microhardness measurements, and *in situ* during heating, by high precision electrical resistance measurements. Both materials show stable UFG structure up to 440°C. Further annealing causes recovery and recrystallization of the microstructure. At 650°C, the microstructure is completely recrystallized. Magnesium alloy AZ31 is prepared by hot extrusion followed by ECAP. UFG microstructure recovers and continuously recrystallizes during annealing. The microstructure of UFG AZ31 alloy is stable up to 170°C and subsequent grain growth is analyzed. Special attention is paid to interpret the activation energy of the grain growth. The superplastic properties of UFG AZ31 alloy are investigated in the temperature range of 170–250°C.

Keywords: equal-channel angular pressing, magnesium alloys, titanium alloys, thermal stability, grain growth

1. Introduction

Lightweight metallic (especially titanium and magnesium) materials are extensively used in transport industry and in cutting-edge applications such as manufacturing of medical implants and devices. Ultra-fine grained (UFG) counterparts are still materials of the future, though the first commercial applications are emerging. In some applications, the employed materials are exposed to elevated temperatures either during service or during products manufacturing. Mechanical properties enhancement of materials prepared by severe plastic deformation (SPD) might be reduced due to recovery and recrystallization of the UFG microstructure at higher temperatures.

Magnesium alloys belong to materials with potential to replace aluminum or some other conventional structural materials in automotive, aircraft, and other industry branches. Magnesium and its alloys are light metals with relatively good mechanical properties which provide expanding potential in weight-critical applications. Interest in magnesium-based metals has recently been revived primarily due to their gradually decreasing costs and the effort of scientists, researchers, and engineers to cut down energy consumption and greenhouse gas emissions [1].

Utilization of titanium and titanium alloys for load-bearing orthopedic implants of big joints and for dental implants still increases [2–4]. Advantages of these materials include extreme corrosion resistance, sufficient biocompatibility, moderate elastic modulus, etc. [5]. A material with enhanced strength is required to reduce the size of the load-bearing orthopedic and dental implants. Vast majority of high-strength β-Ti alloys developed for an aircraft industry are not utilizable in biomedicine because of high content of toxic elements, such as vanadium [6, 7].

Ti-6Al-7Nb alloy was developed as a biocompatible alternative to the most used Ti-6Al-4V alloy. It belongs to $\alpha+\beta$ alloys which contain both α and β phases at ambient temperature. The β-transus temperature of this alloy is 1010°C [8]. UFG microstructure of the studied alloy has been already investigated in Ref. [9], and superior mechanical properties of UFG material were reported [10].

Promising possibility for strength and fatigue performance improvements is the manufacturing of materials with sub-micrometer or even nanoscale grain sizes using SPD techniques [11, 12]. These methods are very efficient in achieving significant grain refinement in polycrystalline materials. UFG materials have usually excellent mechanical properties including high strength and, if the UFG microstructure is sufficiently stable, a superplastic capability at elevated temperatures [11, 13]. Nowadays, the most attractive SPD techniques are equal-channel angular pressing (ECAP) [14] or combined process of extrusion followed by ECAP (EX-ECAP) [15], high-pressure torsion (HPT) [16], and accumulative roll-bonding [17]. In practice, ECAP or EX-ECAP processes are especially useful because of their simplicity in laboratory operation. Moreover, these techniques can produce relatively large billets for industrial applications [18]. There are several reports to date of the successful processing of magnesium and titanium materials using ECAP at elevated temperatures by employing different processing procedures [19–26].

The practical applications of the UFG AZ31 magnesium alloy are limited due to a low microstructure stability at elevated temperatures that complicates the processing of final products in industry. Thermal stability depends on many variables, such as stacking fault energy of the material, processing or volume fraction of grain boundaries, and their properties [27]. Microstructure stability at elevated temperature can be improved by various alloying elements or composite reinforcements. Microstructure stability of the AZ31 magnesium alloy after ECAP was studied by Kim [28] or Radi and Mahmudi [29], who investigated the AZ31 alloy reinforced by alumina nanoparticles. Both papers present calculations of grain growth activation energies which identified two or three temperature regimes with significantly different values of activation energy.

The main objective of this work is the investigation of microstructure stability during annealing of the UFG materials, in particular of the AZ31 magnesium alloy, commercially pure titanium (CP Ti) and Ti-6Al-7Nb alloy prepared by ECAP.

2. Materials and methods

As cast commercial AZ31 magnesium alloy (nominal composition of Mg-3%Al-1%Zn) was extruded at 350°C with an extrusion ratio of 22; subsequently, it was processed by four passes of ECAP. ECAP pressing was performed at 180°C with the velocity of 50 mm/min following route B_C, i.e. rotating the sample by 90° between the individual passes. The angle between two intersecting channels and the corner angle were $\Phi = 90°$ and $\Psi = 0°$, respectively. Both channels had a square cross section of 10 × 10 mm. The ECAP die was equipped with an ejector that allows pushing the sample out of the die immediately after pressing from the feed-in channel to the exit channel.

Flat squared specimens were cut from the middle part of the billets perpendicular to the pressing direction. Results of the microstructural observations and microhardness measurements on planes parallel to the pressing direction are very similar to those from the perpendicular plane [30] and were not addressed in this work.

CP Ti Grade 4 [31] was processed by ECAP through die with the round channel having the diameter of 15 mm. The channel angle was $\Phi = 105°$ and the corner angle $\Psi = 20°$. The temperature of the die during pressing was 300°C. The billets were pressed six times (six passes) following the route B_C at a constant ram speed of 60 mm/min. A detailed study of the ECAP processing of CP Ti can be found in Ref. [32]. Benchmark coarse grained material was prepared by annealing of the as-received material at 800°C for 2 h followed by slow cooling in furnace.

Ti-6Al-7Nb alloy was prepared by multi-step thermal sequence before processing by ECAP. The thermal treatment consisted of two subsequent annealing steps. The first annealing was at 985°C (a temperature just below β-transus) for 1 h and the second annealing at 780°C for 4 h. Each thermal treatment was followed by water quenching. The annealed material possesses a common "duplex" structure, which consists of 18 vol.% of primary α-phase [31]. Such microstructural condition allowed the successful material processing by ECAP. The ECAP die with round channel with the diameter of 20 mm and angles of $\Phi = 120°$ and $\Psi = 0°$ was used for processing. The samples were pressed six times (six passes) at the temperature of 600°C. Subsequently, extrusion to 10 mm at 300°C was applied. Finally, the material was aged at 500°C for 1 h to achieve the maximum strength level. More details about material and its processing can be found in Ref. [33]. Benchmark coarse grained material underwent the same thermal treatment including the annealing steps simulating the thermal history during ECAP and extrusion.

The samples of CP Ti and Ti-6Al-7Nb alloys were heated up to the three temperatures specified by *in-situ* electrical resistance measurements (described in detail below) and subsequently water quenched.

Specimens of AZ31 magnesium alloy for thermal stability investigation were prepared by isochronal annealing for 1 hour at the temperatures ranging from 150 to 500°C followed by water quenching. Specimens of AZ31 magnesium alloy were mechanically grinded on watered abrasive papers and then polished with polishing diamond suspensions of grade 3, 1 and ¼ μm. Flat samples for Vickers microhardness measurements (load 100 g, 10 s) with minimum surface scratches were obtained by this method. Finally, the specimen's surface was polished by argon ions (Gatan PIPS™), which guaranteed successful electron backscatter diffraction (EBSD) measurements. CP Ti and Ti-6Al-7Nb alloy were prepared by mechanical grinding and polishing using watered abrasive papers followed by three-step vibratory polishing. The polished samples were prepared for microhardness (load 500 g, 10 s) and SEM measurement.

For *in-situ* measurement of electrical resistance evolution during heating, a precise self-made apparatus utilizing a common four-point method was employed. The electrical current and voltage were measured simultaneously. The samples were placed in a specially designed furnace which allows precise heating of the sample in a protective argon atmosphere. The relative error of such measurement is lower than 10^{-4}, and the experimental values are acquired with the frequency of 2 Hz [34]. The dynamics of microstructural changes can be assessed from these measurements. The electrical resistance was measured during heating with the constant rate of 5°C/min up to 700 and 800°C for CP Ti and Ti-6Al-7Nb alloy, respectively. UFG conditions of both materials were investigated along with their annealed coarse grained (non-deformed) counterparts. The samples for these measurements require special design maximizing their effective length.

Microhardness of AZ31 alloy and CP Ti was measured by LECOM-400-A microhardness tester. For Ti-6Al-7Nb alloy, a QNESS A10+ microhardness tester was employed with automatic indentation and evaluation using the QPix Control Program. FEI Quanta 200 FX scanning electron microscope equipped with EDAX EBSD camera and OIM software was utilized for EBSD and microstructure observations.

Flat specimens of AZ31 magnesium alloy for tensile tests were machined and cut from ECAPed billets parallel to extrusion direction. The continuous measurements of *m*-parameter were performed on six samples from a single ECAPed billet. The gauge length was 16 mm, and the thickness and width were approximately 1 and 4 mm, respectively. Tensile tests were performed using a screw-driven Instron 5882 machine at 175, 200, and 250°C. Computer-operated machine allows arbitrary control of cross-bar movement.

Atomic force microscopy (AFM) observations were performed to study the deformation mechanism. The tensile samples were carefully polished on grinding papers and using diamond pastes (3, 1, and ¼ μm) before the tensile test. The samples after deformation were observed using Bruker Dimension Edge AFM.

3. Results

3.1. *In-situ* electrical resistance measurements of CP Ti and Ti-6Al-7Nb alloy

Temperature dependence of electrical resistance of CP Ti after ECAP and of the annealed coarse grained material is shown in **Figure 1(a)**. The relative resistance $R(T)/R_0$, where $R(T)$ is the resistance measured as the function of temperature T and R_0 is the resistance at room temperature, is plotted at the vertical axis. During heating up to 700°C, the resistance increases almost three times. Initially, the resistance increases linearly, whereas above 300°C, the evolution with temperature becomes concave. Small difference between annealed and ECAPed

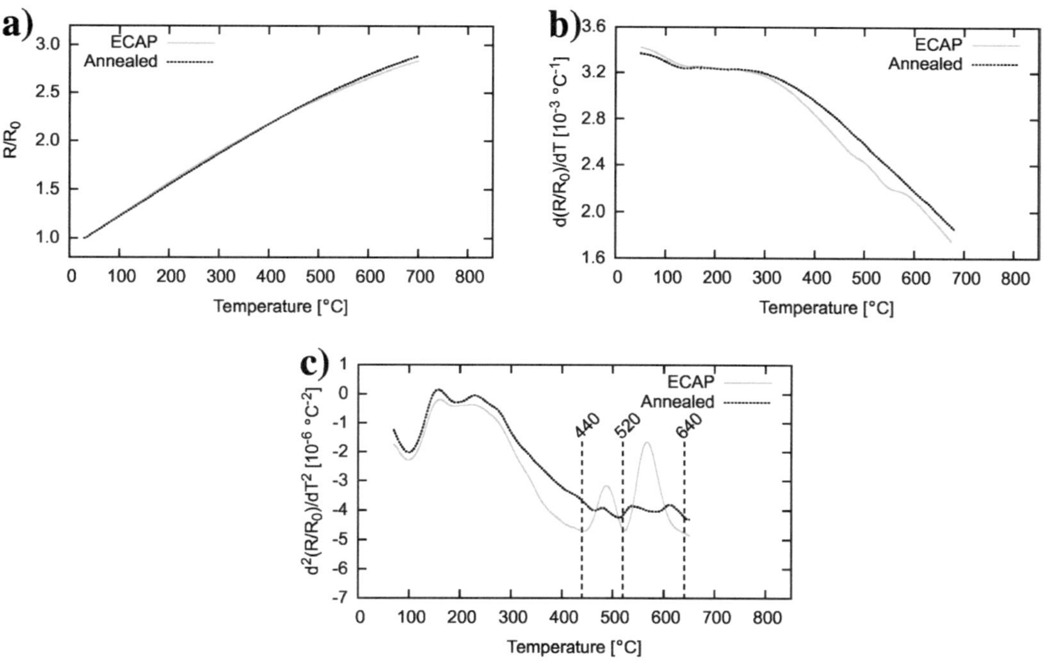

Figure 1. Relative resistance variations of CP Ti during heating (a) temperature dependence, (b) the first derivative of relative resistance, and (c) the second derivative of relative resistance (highlighted temperatures were chosen for SEM observation).

samples is observed. The first and the second derivatives of relative resistance with respect to temperature, computed numerically, are shown in **Figure 1(b)** and **1(c)**, respectively. In **Figure 1(b)**, both curves behave in a similar manner with two small deviations for ECAPed specimen at about 500 and 600°C. Those deviations appear as well observable peaks in the plot of the second derivative, which is plotted in **Figure 1(c)**. Assuming that these peaks correspond to undergoing microstructural changes, the temperatures of 440, 520, and 640°C were chosen for subsequent annealing and *ex-situ* observations of microstructure. All other peaks in the second derivative graph appear in both curves and therefore they do not represent differences caused by different initial microstructure.

Figure 2 shows the results of the resistance measurements of Ti-6Al-7Nb alloy that are presented in the similar way as for CP Ti. **Figure 2(a)** shows the temperature dependence of the relative resistance for the UFG Ti-6Al-7Nb alloy after ECAP and in the as-rolled condition. The relative resistance increases only by approximately 10%, in contrast to the CP Ti. The difference of the ECAP and as-rolled condition is therefore relatively more pronounced in **Figure 2(a)**. The overall course of both curves is concaved up to 650°C, and for higher temperatures, the electrical resistance even declines. **Figure 2(b)** shows the first derivative of electric resistance with respect to the temperature with two distinct peaks around 500 and 650°C for the ECAPed material. The differences between the two conditions are

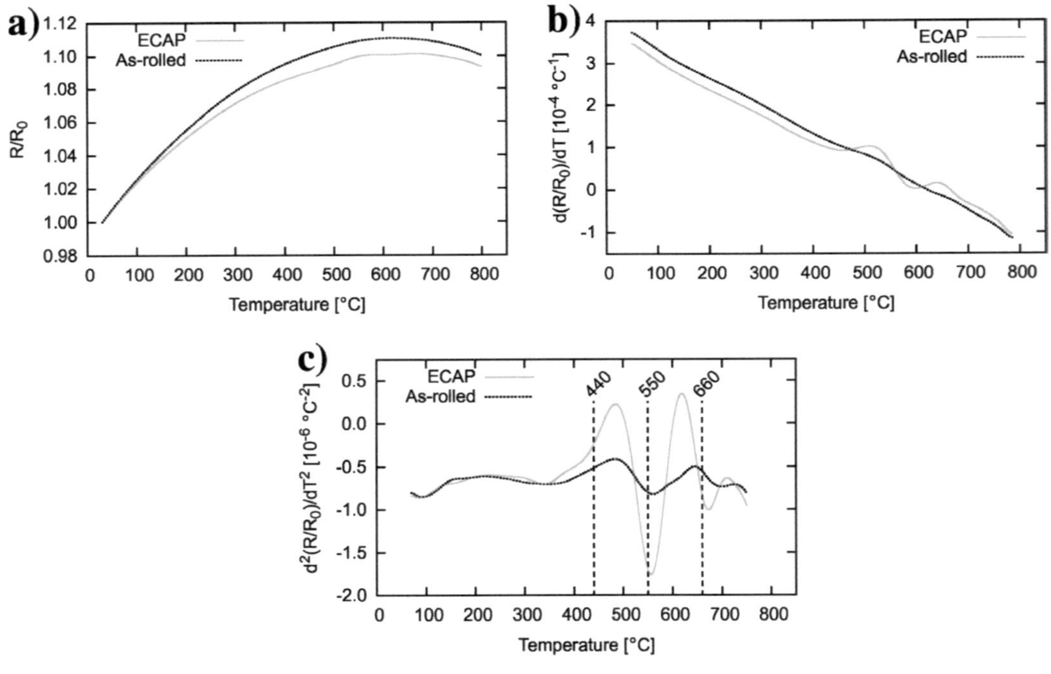

Figure 2. Relative resistance variations of Ti-6Al-7Nb alloy during heating (a) temperature dependence, (b) the first derivative of relative resistance, and (c) the second derivative of relative resistance (highlighted temperatures were chosen for SEM observation).

accentuated by plotting the second derivative of the electrical resistance as displayed in **Figure 2(c)**. The temperatures of 440, 550, and 660°C were selected for the microstructure observations using SEM.

3.2. Mechanical properties

The microhardness of the UFG materials after the SPD processing ($HV0.1_{AZ31} = 86$, $HV0.5_{CPTi} = 274$, and $HV0.5_{Ti67} = 369$) is significantly higher than that of the annealed conditions ($HV0.1_{AZ31} = 58$, $HV0.5_{CPTi} = 215$, and $HV0.5_{Ti67} = 283$) [31, 35, 36]. **Figure 3** depicts the microhardness dependence on the annealing temperature for each sample.

Microhardness values of the AZ31 alloy after annealing at 150 and 170°C do not differ significantly. However, the microhardness of the AZ31 declines abruptly in the temperature range of 170–230°C and then continues to decrease up to 500°C.

The values of microhardness of CP Ti remain nearly constant ($HV0.5 \approx 280$) up to the aging temperature of 450°C and then decrease rapidly approaching values of annealed material at 700°C ($HV0.5_{annealed} = 215$).

The annealed sample of Ti-6Al-7Nb exhibits the microhardness of $HV0.5_{annealed} = 283$. The microhardness of ECAPed specimen increases up to ($HV0.5 \approx 370$) and remains almost constant after heating to 440 and 550°C. Only heating to the highest temperature (660°C) results in a slight decrease of HV. However, the decrease of HV is much lower than in other investigated materials.

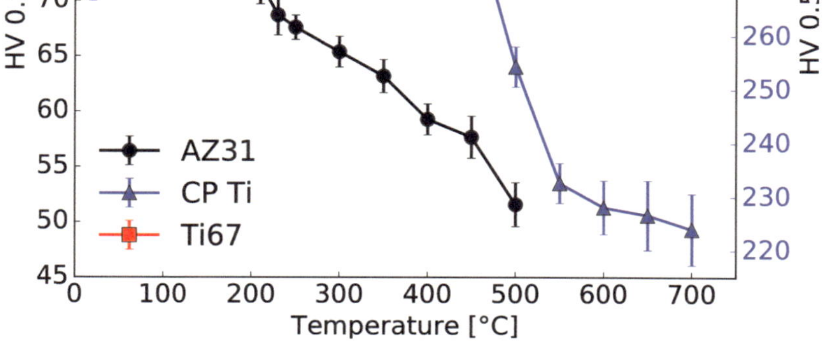

Figure 3. Microhardness of the studied materials subjected to SPD and heat treatment.

3.3. Microstructure

3.3.1. Microstructure changes and dislocation density evolution of UFG AZ31 alloy during heating

The microstructure of AZ31 magnesium alloy after extrusion (not shown here) is bimodal containing large grains elongated in the extrusion direction (≈10 μm) and smaller grains (≈1 μm) [36].

UFG microstructure of the specimen in the initial non-annealed condition (after extrusion and four passes of ECAP) is shown in **Figure 4(a)**. The microstructure is homogeneous comprising fine grains of the average size of 0.9 μm. The microstructure and average grain sizes of the samples after 1 h of isochronal annealing at 150 and 170°C (not shown here) are similar to the initial non-annealed specimen.

Inhomogeneous grain growth is observed at higher annealing temperatures (**Figure 4(b)–(f)**). Some grains start to grow at annealing temperatures of 190 and 210°**C** (the microstructure of the sample after annealing at 210°**C** is similar to that of 190°**C** and is not shown here). The fraction of coarse grains increases with increasing annealing temperature. At annealing temperature of 250°**C**, some areas with original fine grains are still observed. However, the small grains (≈1 μm) are continuously disappearing at higher annealing temperatures, and nearly no small grains are observed after annealing at 400°**C** (see **Figure 4(e)**). Please note that

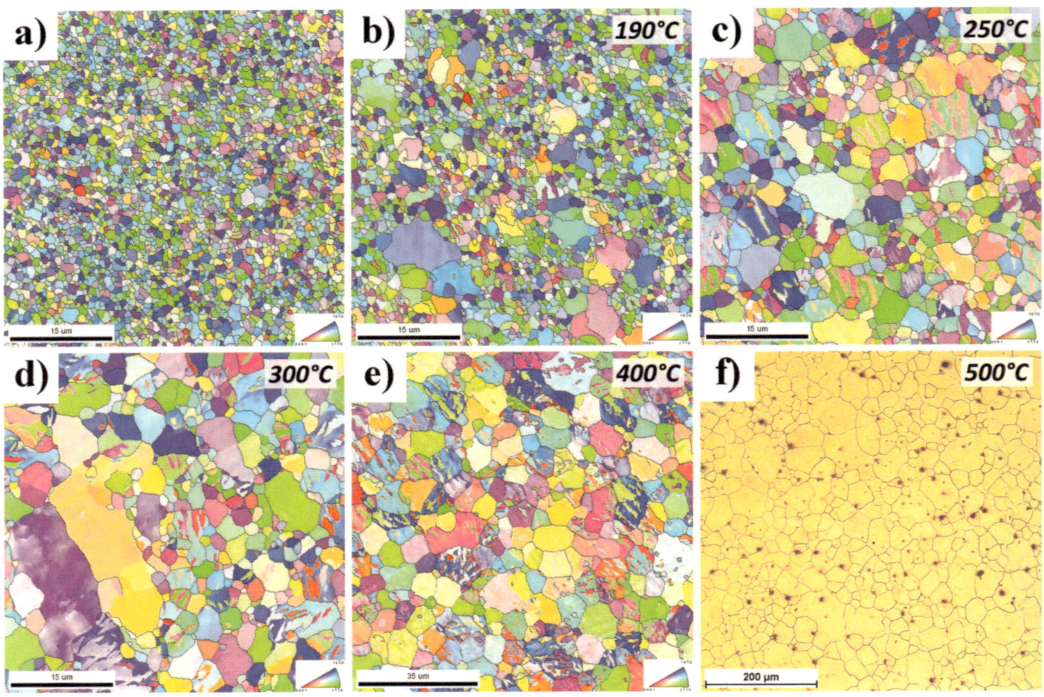

Figure 4. Microstructure of the AZ31 sample after extrusion and four passes of ECAP (a), and isochronally annealed at (b) 190°C, (c) 250°C, (d) 300°C, (e) 400°C, and (f) 500°C. Images (a)–(e) show results from EBSD measurements. The image (f) was taken using light microscope.

http://dx.doi.org/10.5772/intechopen.68956

the magnification of the EBSD image in **Figure 4(e)** is two times smaller than the magnification of the previous EBSD images. Microstructure of the specimens annealed at 500°C was observed by light microscope (see **Figure 4(f)**).

The dependence of average grain sizes (number average) on annealing temperatures is plotted in **Figure 5(a)**. In specimens annealed at 250 and 300°C, the average values are calculated from the bimodal grain size distribution. The dependence of the average grain sizes and microhardness values on annealing temperature is summarized in **Table 1**.

Annealing twins observed after annealing at 250–400°C (see **Figure 4(c)–(e)**) were excluded from grain size calculations to achieve the true grain size values. All these twins were identified as the tensile twins with the misorientation angle of 86° [37].

The plastic shear deformation by extrusion and ECAP causes the accumulation of large plastic strain and the increase of density of structural defects. These defects are stable at room temperature, but they annihilate relatively easily during annealing.

The dependence of the mean dislocation density ρ_D measured by positron annihilation spectroscopy (PAS) for the samples subjected to annealing treatment at various temperatures is shown in **Figure 5(b)**. Dislocation density decreases with increasing annealing temperature and falls below the detection limit of PAS at annealing temperatures $T \geq 300°C$.

3.3.2. *Microstructure changes of UFG CP Ti and Ti-6Al-7Nb alloy during heating*

Microstructure changes in the UFG CP Ti and Ti-6Al-7Nb after ECAP occurring during linear heating were investigated *ex situ* by SEM. Samples in conditions corresponding to linear heating to the temperatures selected from *in-situ* electrical resistance measurements were observed (440, 520, and 640°C for CP Ti; 440, 550, and 660°C for Ti-6Al-7Nb alloy).

Figure 6 shows the microstructure of CP Ti, while in **Figure 6(a)**, the UFG microstructure of material after ECAP is displayed. White dots in the SEM micrograph are β-Ti particles formed due to contamination by iron, which is typical for CP Ti. High Fe content in these particles was proved by energy dispersive X-ray spectroscopy. The microstructure of the material is typical

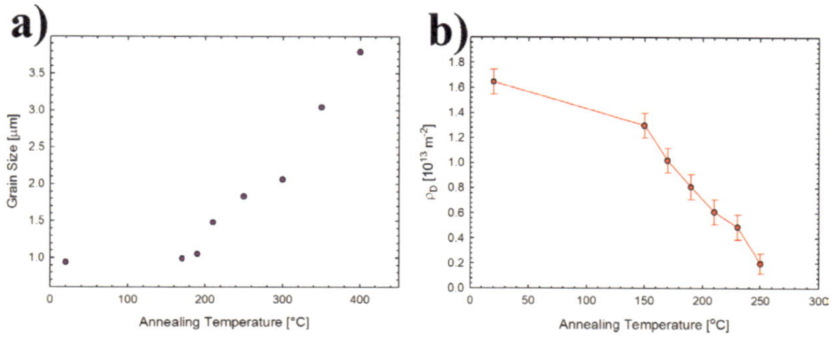

Figure 5. (a) Dependence of the average grain size (number average, excluding twins) of the AZ31 alloy on annealing temperature (up to 400°C). (b) Dependence of the dislocation density on annealing temperature after 1 h of isochronal annealing process.

Annealing temperature [°C]	–	170	190	210	250	300	350	400	450	500
Microhardness HV0.1	85.8	84.1	78.0	71.6	67.6	65.4	63.2	59.3	57.7	51.6
Average grain size *d* [mm]	0.94	0.99	1.05	1.48	1.83	2.06	3.04	3.79	10.09	24.53

Table 1. Microhardness values and average grain sizes at different annealing temperatures.

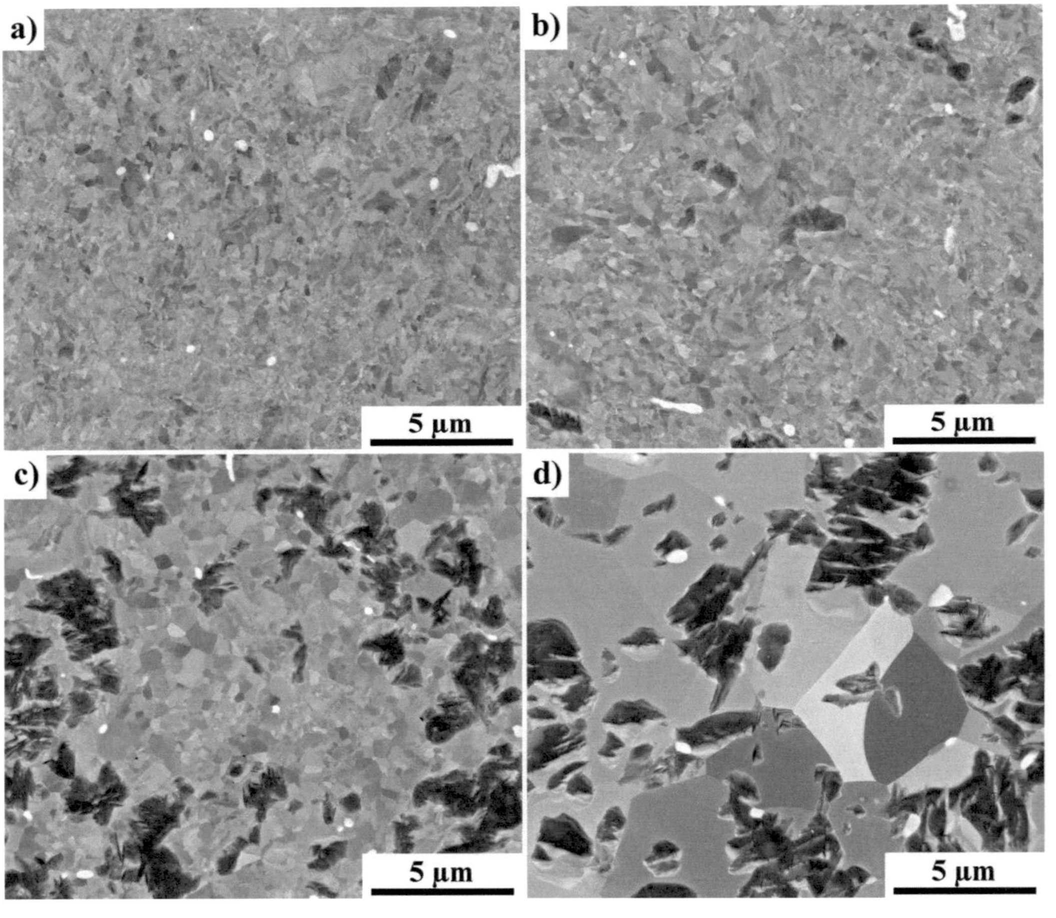

Figure 6. Microstructure evolution of ultra-fine grained CP Ti (a) as ECAPed, (b) heated to 440°C, (c) heated to 520°C, and (d) heated to 640°C.

heavily deformed containing grains with the average size around 1 μm [11, 38]. No significant differences of the microstructure were observed in the sample annealed up to 440°C (see **Figure 6(b)**). On the other hand, the microstructure of the sample annealed to 520°C differs considerably as can be seen in **Figure 6(c)**. The grains are much clearer, which suggests that some recovery process, probably annihilation of dislocations, was undergoing during heating between 440 and 520°C. Grain size also slightly increased. The dark spots in the micrograph are probably artifacts caused by polishing. The microstructure of the specimen heated up to 640°C is shown in **Figure 6(d)**. Material is completely recrystallized with grains of the average size of approximately 5 μm.

Figure 7 shows the microstructure of UFG Ti-6Al-7Nb alloy after ECAP and subsequent heating. The material after ECAP, as shown in **Figure 7(a)**, has the typical duplex microstructure consisting of approximately 20% of heavily deformed primary α-phase and significantly fragmented α + β region, which contains slightly elongated β-phase particles appearing white in the micrograph due to chemical contrast. The microstructure of ECAPed specimen subsequently heated up to 440°**C** as shown in **Figure 7(b)**. There are no observable changes in the microstructure as compared to the ECAPed specimen. **Figure 7(c)** displays the material annealed up to 550°**C**. Detailed inspection of the micrograph reveals small fraction of tiny grains in α + β region with very clear contrast, suggesting that these are newly formed dislo-

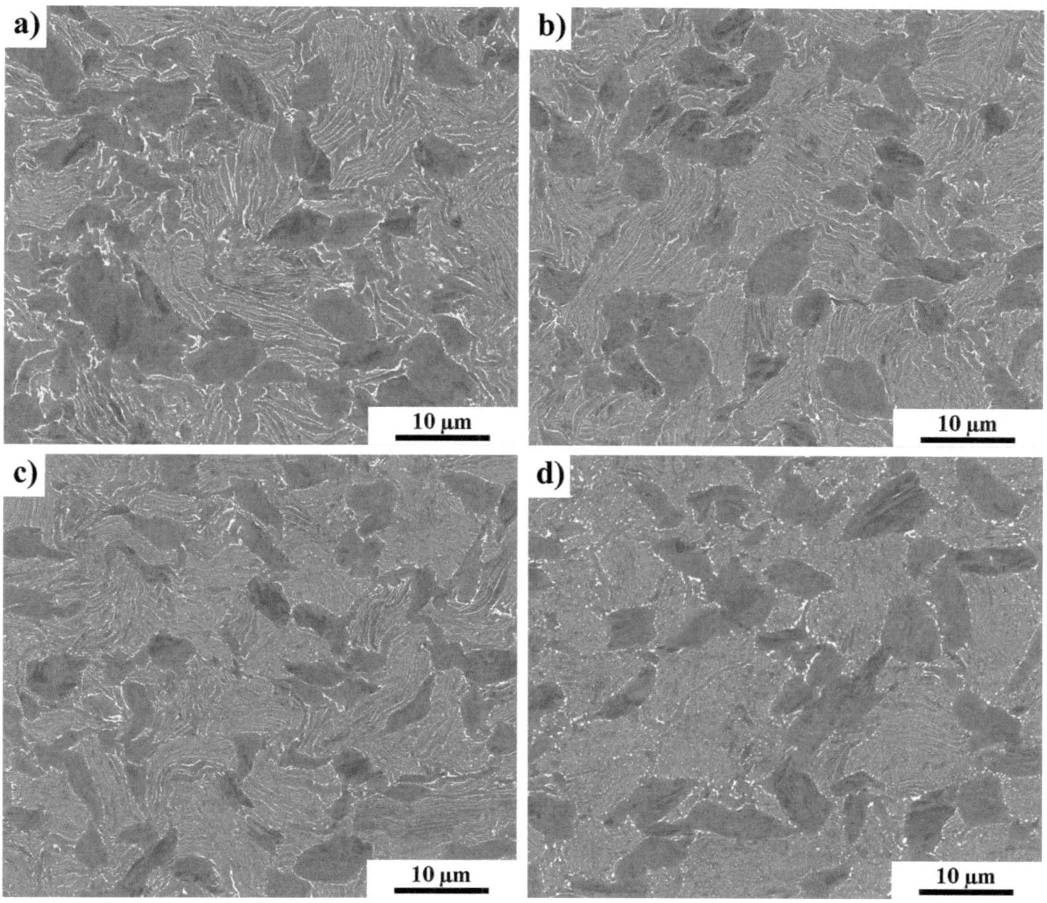

Figure 7. Microstructure evolution of UFG Ti-6Al-7Nb alloy (a) as ECAPed, (b) heated to 440°C, (c) heated to 550°C, and (d) heated to 660°C.

cation-free grains. Also β-phase particles are slightly globularized. Finally, in **Figure 7(d)**, the microstructure of the specimen annealed up to 660°**C** is shown. The microstructure is partly recrystallized with grains >1 μm in the originally heavily fragmented α + β region. White β-phase particles are significantly bigger and more globular.

3.4. Superplastic behavior of AZ31

3.4.1. Methodology of superplastic behavior determination

Two types of tests for strain-rate sensitivity determination were performed. Firstly, standard strain-rate changes tests were carried out to determine the *m*-parameter for a wide range of strain rates at selected temperature. True strain rate was increased in a step-wise manner from 5×10^{-5} to 10^{-2} s^{-1}. Maximum stress after approximately 2% deformation at each strain rate was recorded for the calculation of *m*-parameter.

Secondly, a special strain rate control test was undertaken. For the determination of the strain rate sensitivity at strain rate of $\dot{\varepsilon}_1 \approx 10^{-4} s^{-1}$, two different true strain rates were selected: $\dot{\varepsilon}_1 = 0.9 \times 10^{-4} s^{-1}$ and $\dot{\varepsilon}_2 = 1.2 \times 10^{-4} s^{-1}$. The actual true strain rate was changed every 120 s (i.e. after $\varepsilon \approx 1.2\%$) during the experiment. Note that the overall cross-bar speed exponentially increased to maintain the selected two true strain rates. Therefore, the overall true strain is proportional to the time ($\varepsilon = 100\% \sim t = 3h$). The methodology of continuous measurement of *m*-parameter during the tensile test is described in detail in our recently published paper [39].

Due to the strain rate sensitivity of the material and alternating strain, the resulting flow curve (thin curve in **Figure 8** for sample deformed at 200°**C**) has a saw-like character.

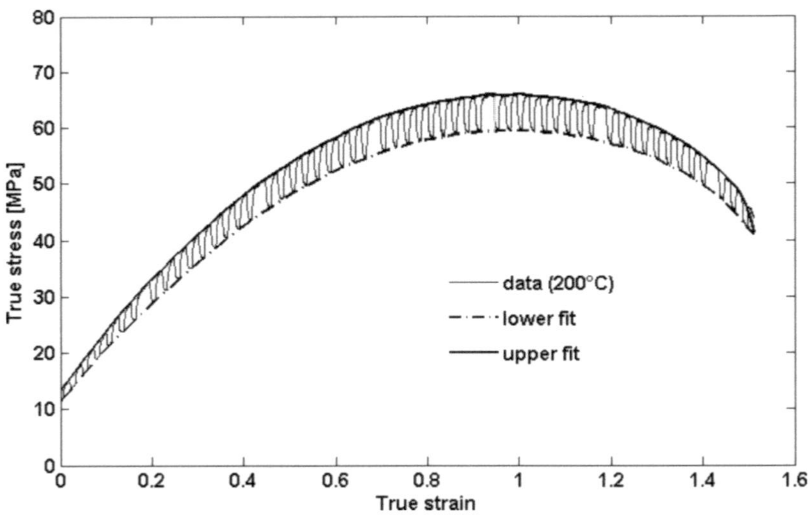

Figure 8. Measured flow curve for alternating strain rates (thin curve) and flow curves interpolated through local maxima and minima (thick smooth curves) for sample deformed at 200°C.

Local maxima of the alternating flow-curve were joined by a smooth curve regarded to as the "upper fit" which represents the approximate flow-curve at the higher strain rate $\sigma_2(\varepsilon)$, whereas the interpolation of local minima, the "lower fit", estimates the flow-curve at the lower strain rate $\sigma_1(\varepsilon)$. As a result, the continuous evolution of m-parameter with strain can be calculated as:

$$m(\varepsilon) = \frac{ln(\sigma_2(\varepsilon)) - ln(\sigma_1(\varepsilon))}{ln(\dot{\varepsilon}_2) - ln(\dot{\varepsilon}_1)}. \quad (1)$$

Note that the denominator in Eq. (1) depends only on the selected true strain rates and is constant.

3.4.2. Results: superplastic behavior of UFG AZ31 alloy

The evolution of m-parameter with strain rate is depicted in **Figure 9** for testing temperatures of 175, 200, and 250°C. Material exhibits the superplastic behavior ($m > 0.5$) at all studied temperatures and strain rates up to 10^{-4} s^{-1}. At intermediate temperatures of 175 and 200°C, the range of $m > 0.5$ extends to strain rates of 5×10^{-4} s^{-1}. For strain rates higher than 10^{-3} s^{-1}, the material is not superplastic ($m < 0.3$) at all studied temperatures.

Based on these results, the strain rate of 10^{-4} s^{-1} and temperatures of 175, 200, and 250°C were selected for further testing employing alternating strain rate, as described in the previous section. Two samples per condition were tested. **Figure 10** shows the measured true stress-true strain flow curves. Both measured flow curves for each condition are shown to assess the reproducibility of the experiment. All flow curves exhibit significant strain hardening,

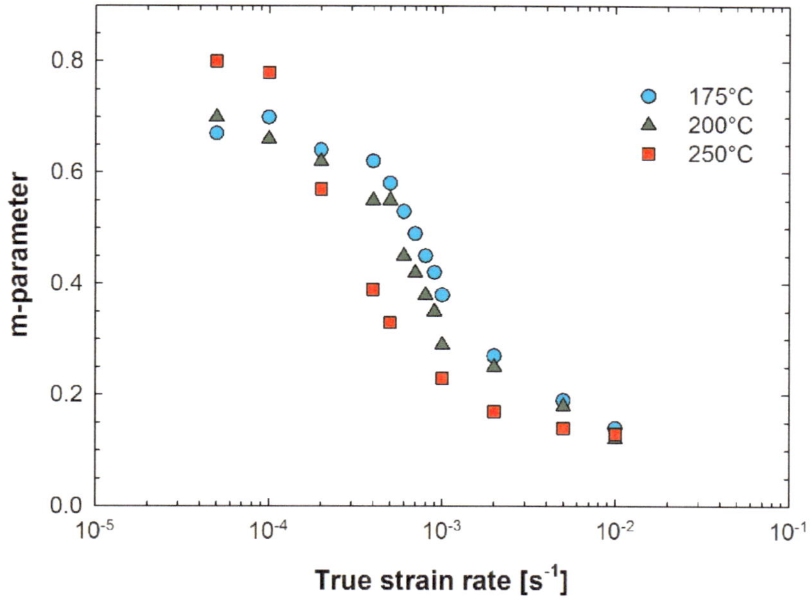

Figure 9. Measured m-parameter at deformation temperatures 175, 200, and 250°C.

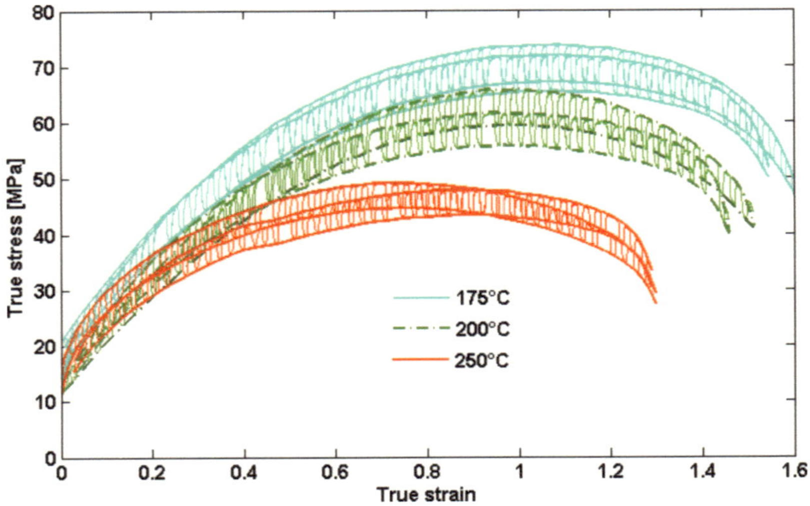

Figure 10. True stress-true strain curves for six tested samples. Measured flow curve for alternating strain rates (thin curve) and flow curves interpolated through local maxima and minima (thick smooth curves).

which is followed by moderate softening. In the final stage of deformation, observed softening is much more pronounced and the strain rate sensitivity decreases, which suggests that the specimen undergoes the strain localization. As expected, the highest stress is achieved at the lowest testing temperature of 175°**C**. However, the samples tested at 175°**C** exhibited surprisingly the highest elongation to fracture ≈ 380% (true strain $\varepsilon \approx 157\%$). The flow curves for 200°**C**, and especially 250°**C**, reached the lower maximum true stress, which was also achieved at lower true strain. Shorter range of strain hardening seems to be responsible for lower achieved total elongation, especially in samples deformed at 250°**C**.

The m-parameter evolution $\boldsymbol{m(\varepsilon)}$ calculated from Eq. (1) for all investigated samples is depicted in **Figure 11** [39]. In the beginning of test, the m-parameter reaches 0.5 and then decreases with increasing true strain to values slightly above 0.3. The m-parameter for samples deformed at 250°**C** is lower in the initial stage of the deformation, while the m-parameter for samples deformed at 200°**C** is the highest at the true strain ε **> 1**. Final sharp decrease of m-parameter is associated with necking.

Achieved elongation and m-parameter values suggest superplastic deformation mediated by grain boundary sliding. If a sample with polished smooth surface is deformed in superplastic regime by grain boundary sliding, individual grains can be observed on surface using atomic force microscopy (AFM) [40–42]. Tensile sample deformed at 150°**C** with the constant strain rate of 10^{-4} s^{-1} which achieved elongation of 315% was used for AFM measurement. **Figure 12(a)** shows the deformed region far from the neck. This region was deformed superplastically, and individual grains with the size of ~1 μm can be observed. On the other hand, **Figure 12(b)** shows the region close to the tip of the neck, where the failure occurred. Slip bands appear as typical steps and grain structure cannot be resolved.

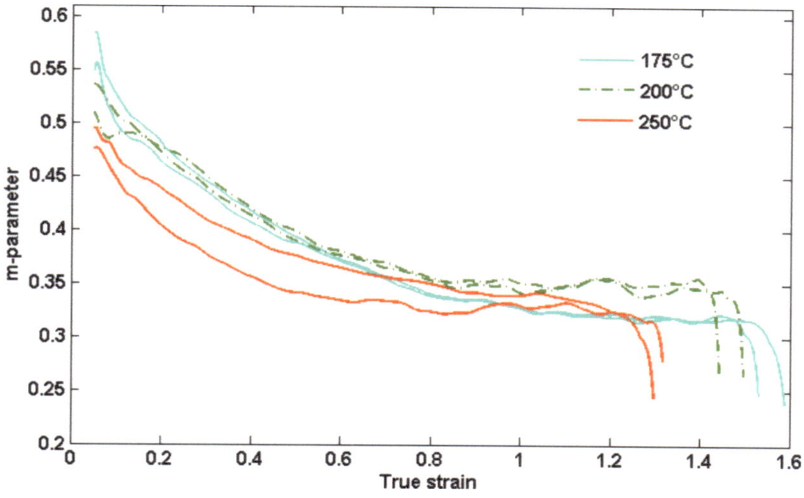

Figure 11. Evolution of m-parameter determined from interpolated flow curves for samples deformed at 175, 200, and 250°C.

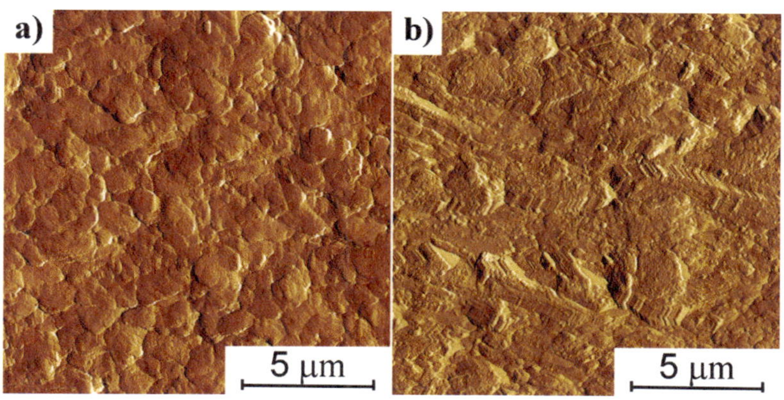

Figure 12. AFM image of a surface after deformation: (a) deformed region far from the neck, (b) deformed region in the neck.

4. Discussion

4.1. AZ31 magnesium alloy

4.1.1. Correlation between mechanical properties, dislocation density, and microstructure

Microhardness measurements (cf. **Figure 3**) indicate that UFG microstructure of AZ31 magnesium alloy is stable up to 170°C. After annealing at temperatures higher than 190°C, a sharp drop of microhardness occurred. A detailed inspection of the temperature dependence of the microhardness (cf. **Figure 3**) indicates a two-step character of the microhardness decline. In the lower annealing temperature range (170–210°C), the decline is significantly sharper, while for higher annealing temperatures ($T > 210$°C), the slope of the curve is much lower.

This two-step character of the curve suggests a change of the mechanism controlling mechanical properties. The strength and microhardness of severely deformed UFG materials are affected mainly by the dislocation density [43] and the grain size according to the Hall-Petch relation [44, 45]. Therefore, the grain coarsening and the dislocation annihilation during annealing are expected to control the material strength and microhardness.

Figure 5(a) shows the grain sizes evolution, which could be correlated with dislocation density evolution with annealing temperature shown in **Figure 5(b)**. In the low temperature region of the microhardness drop ($T \approx 170$–210°C), the grain growth is relatively negligible (see **Table 1**), whereas the dislocation density gradually declines indicating a recovery of dislocation structure. Most probably rearrangement and mutual annihilation of dislocations with opposite signs take place during annealing in this lower temperature range ($T < 210$°C). As seen in **Figure 4(b)** and **(c)**, the fine grain structure becomes unstable and significant grain growth is observed at temperatures $T > 210$°C. In this temperature range, the dislocation density is very low, falling below the detection limit of PAS ($\rho_D \approx 10^{12}$ m^{-2}) at $T \approx 300$°C [36].

From microstructure observation (using EBSD) and lattice defect density determination (using PAS), one can conclude that in the lower annealing temperature region ($T \approx 180$–210°C), it is mostly the annihilation of dislocations which causes the drop of microhardness. At higher annealing temperatures ($T > 210$°C), probably the grain growth influences significantly the hardness of AZ31 magnesium alloy.

4.1.2. Grain growth analysis

The determination of grain size in UFG material allows analyzing the mechanisms of grain growth during annealing. Two microstructural aspects may be determined:

(a) The activation energy of grain growth

The grain growth mechanism during static annealing can be assessed from calculated activation energy of grain growth. For this analysis, we can use the general equation of the grain growth

$$d^n - d_0^n = kt, \tag{2}$$

where d is the grain size after given annealing time, d_0 is the initial grain size, n is the grain growth exponent, t is the annealing time, and k is a temperature-dependent constant which can be described by Arrhenius equation:

$$k = k_0 \exp\left(-\frac{Q}{RT}\right), \tag{3}$$

http://dx.doi.org/10.5772/intechopen.68956

where k_0 is a constant, Q is the activation energy of grain growth, R is the universal gas constant, and T is the thermodynamic temperature.

The value of the stress exponent n is of significant importance. In the ideal case (defect-free infinite crystal), the grain growth exponent n should be equal to 2. However, higher values of n are very often found, which can be attributed to various factors affecting grain growth kinetics, such as the effect of free surface, texture, impurity-drag, dislocation substructure, and microstructure heterogeneities [46]. Several studies [47–49] reported a value of n in a range from 2 to 8 for various magnesium alloys and magnesium-based composites. Higher values of n ($n \geq 5$) were observed mainly in UFG magnesium materials produced by mechanical alloying [48, 49]. The value of grain growth exponent n observed in ultra-fine grained magnesium alloy AZ31 produced by various techniques of severe plastic deformation ranges between 2 and 4 [47, 50, 51]. The AZ31 alloy processed similarly as the investigated material (ECAP without previous hot extrusion, where the average grain size after 4 passes was equal to 2.5 μm) was studied by Kim [28] and Kim and Kim [47]. The grain growth exponent n used in their calculations was equal to 2. We use the same value of n, which will allow us to make a direct comparison with the results of Kim and Kim [47].

Considering isothermal annealing and substituting Eq. (3) into Eq. (2), one can determine the activation energy Q as the slope of the dependence of $ln(d^2 - d_0^2)$ on T^{-1} which is shown in **Figure 13** for the investigated AZ31 alloy. Three temperature ranges with different Q values can be distinguished. The calculated values of activation energy of grain growth are 115, 33,

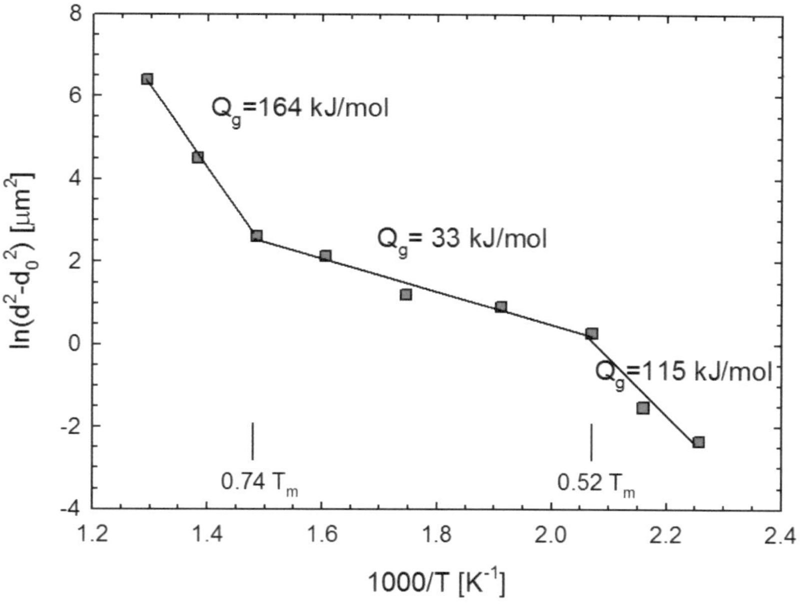

Figure 13. Plot of $ln(d^2 - d_0^2)$ vs. T^{-1} for the estimation of the activation energy of grain growth of the EX-ECAP magnesium alloy AZ31.

and 164 kJ/mol in the temperature ranges 170–210, 210–400, and 400–500°C (443–483, 483–573, and 573–673 K), respectively. These three temperature ranges with different Q values were observed in other fine-grained AZ31 alloys in various conditions, and the respective temperature ranges are very similar with our temperature ranges [39, 47].

In the low temperature range (T< 483 K, <210°C), the activation energy is relatively high—higher than the activation energy of grain boundary diffusion in pure magnesium (92 kJ/mol [52]), but, on the other hand, much lower than the activation energy of lattice self-diffusion (135 kJ/mol [53]). Considering a well-known fact that the activation energy of alloys should be higher than the activation energy of pure metals, the diffusion mechanism can be attributed to the grain boundary diffusion, which might be further affected by dislocations. In this temperature range, the dislocation density within the grains decreases with increasing temperature, but it remains relatively high.

In the high temperature range (T > 673 K, > 400°C), the activation energy Q is equal to 164 kJ/mol, which is higher than lattice self-diffusion in pure magnesium (135 kJ/mol [53]). The lattice self-diffusion is activated, and the grain growth leads eventually to fully-recrystallized structure.

In the intermediate temperature range, the value of Q is abnormally low. Similarly, low value of Q was reported by Wang et al. [54] in the ECAPed Al-Mg alloy annealed at the temperatures $T \leq 275$°C. The authors attribute the extremely low value of Q to the non-recrystallized microstructure with a certain fraction of non-equilibrium grain boundaries. This conclusion is consistent with the concept of reduced activation energy of grain boundary diffusion in UFG materials produced by SPD caused by the ability of the non-equilibrium grain boundaries to provide enhanced atomic mobility [55, 56]. The AZ31 alloy after extrusion and 1 pass of ECAP contains a significant number of non-equilibrium grain boundaries. However, the fraction of non-equilibrium grain boundaries decreases with increasing number of ECAP passes so that nearly no such grain boundaries are observed in more deformed AZ31 alloy [57].

It is shown in Kim and Kim [47] and supported by our results that the low fitted value of apparent activation energy in the intermediate temperature range 210–400°C cannot be substantiated. It is argued in [47] that the mechanism of diffusion that is the driving force for grain growth is continuously changing due to recovery processes and therefore the Arrhenius equation (Eq. (1)) is not valid. Detail computation provided in Ref. [58] shows that if true activation energy of the process responsible for the grain growth continuously rises from the activation energy of grain boundary diffusion (115 kJ/mol) to the activation energy of lattice self-diffusion (164 kJ/mol), then the (wrong) fitting by a single Arrhenius equation indeed results in very low (and physically meaningless) estimate of apparent activation energy (33 kJ/mol). Based on a simple model assuming continuous increase of activation energy [58], it can be concluded that the dominant diffusion process is the grain boundary diffusion up to 210°C, while the lattice self-diffusion is dominant from 400°C. In the intermediate region, the effect of grain boundary diffusion decreases due to undergoing grain growth.

(b) Hall-Petch relation

EBSD analysis allows us to determine the validity of the Hall-Petch relation for isochronally annealed UFG AZ31 alloy in the temperature range up to 400°C. For this analysis, the Hall-Petch relation yields

$$HV = H_0 + K_H d^{-\frac{1}{2}}, \tag{4}$$

where HV is the measured value of microhardness and H_0 and K_H are material constants.

The dependence of HV on d determined from **Figures 3** and **5**, respectively, is plotted in **Figure 14**.

The constants H_0 and K_H may be calculated from the parameters of a linear fit depicted also in **Figure 14**. The best fit was applied only to data corresponding to higher annealing temperatures (from 250 to 500°C) since in this temperature range, only the grain size affects the material hardness as the dislocation density is low (cf. **Figure 5(b)**). At low temperatures, both the reduced grain and the high dislocation density contribute to strengthening as one may assess from **Figures 5(a)** and **(b)**, and the linear fit of microhardness data fails. Data for low annealing temperatures (i.e. high dislocation density conditions) lie clearly above the Hall-Petch fit (the difference is marked by the arrow in **Figure 14**).

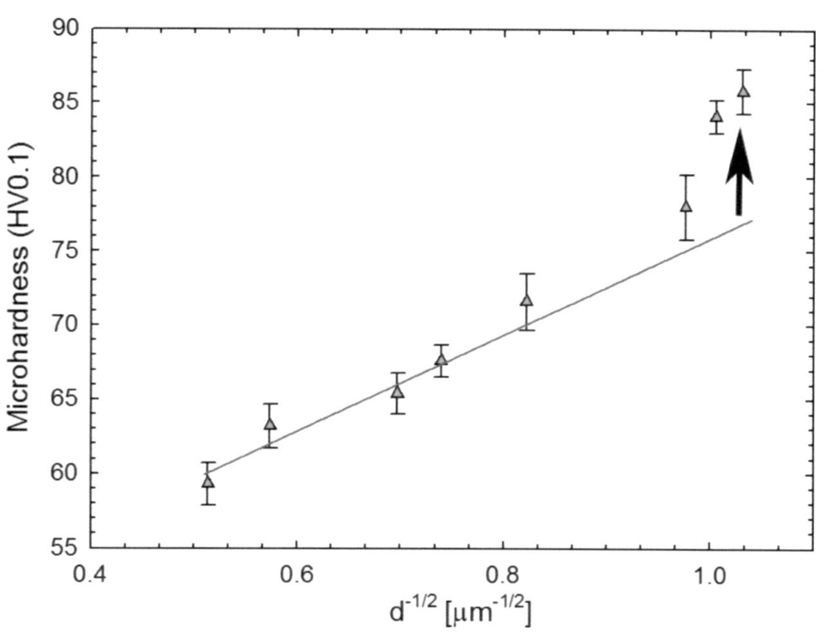

Figure 14. The Hall-Petch relationship for the isochronally annealed EX-ECAP AZ31 alloy based on HV0.1 microhardness data.

The calculated material constants from the high temperature fit of *microhardness* versus $d^{1/2}$ are: $H_0 = 47 \pm 2$ and $K_H = 27 \pm 3$ μm$^{1/2}$. These values are partly comparable to those reported on Al alloys prepared by ECAP (H_0 = 35–47 and K_H = 35–50 μm$^{1/2}$) [59], but different from those reported on the UFG AZ31 alloy reported by Kim and Kim [47] ($H_0 = 38$, $K_H = 42$). It might be argued that the constants in Ref. [47] were calculated from the linear fit of the whole temperature range because the changes of dislocation densities were not taken into account. It results in underestimating and overestimating of H_0 and K_H constants, respectively, in comparison with our calculated values. Our value of the constant H_0 is closer to the microhardness value of the AZ31 in annealed condition (HV0.1 = 58 ± 3, see Ref. [60]) than the value of H_0 calculated by Kim and Kim [47].

4.1.3. *Superplastic behavior*

AZ31 magnesium alloy processed by ECAP exhibited a superplastic behavior at comparatively low temperatures of 150–250°C at low strain rates up to 5×10^{-4} s^{-1} according to *m*-parameter evaluation. This is consistent with previous studies investigating ultra-fine grained AZ31 alloy [61–64]. However, the elongation of studied samples did not reach 400%. This may be partly attributed to the size of used specimens. In this study, we used samples with relatively long gauge length of 16 mm. In this case, final strain localization—necking—is responsible for negligible elongation. On the other hand, if small samples are used (the gauge length ~ 1 mm), necking before failure provides significant additional elongation.

At elevated temperatures, the diffusion processes are generally enhanced and contribute to the superplastic behavior. However, we found that *m*-parameter does not increase with the temperature and achieved elongation even slightly decreases. This unusual behavior can be attributed to recovery and recrystallization processes at elevated temperatures. Diffusion of atoms, which facilitates superplastic behavior, is enhanced by fast diffusion paths like pipe diffusion along dislocations or grain boundaries, which was found as the dominant diffusion processes in severely deformed UFG microstructure [30, 58, 65]. Note also that the activation energy of grain boundary diffusion in pure Mg (92 kJ/mol [52]) is much lower than the activation energy of self-diffusion (135 kJ/mol [53]). The decrease of *m*-parameter and the total elongation for 250°C is therefore probably caused by disappearing of fast diffusion paths due to recovery and grain growth. Limited work hardening at 250°C is also attributed to recovery processes, which occur even during static annealing at 250°C [36, 66], and the grain growth might be even faster under dynamic conditions [67]. On the other hand, during annealing at 175 and 200°C, limited decrease of dislocation density was observed [36]. The *m*-parameter during deformation at 200°C remains higher than at 175°C (possibly due to simple temperature effect on diffusion).

4.2. CP Ti and Ti-6Al-7Nb alloy

4.2.1. *Resistance evolution*

The electrical resistance of CP Ti showed in **Figure 1** increased approximately three times during heating up to 700°C as compared to the room temperature value. In fact, the resistivity increase with increasing temperature in CP Ti depends upon the amount of impurities (mainly oxygen).

The achieved results for CP Ti Grade 4 are in good agreement with other authors [68]. Much smaller increase of resistance (by only 10%) in Ti-6Al-7Nb alloy as compared to CP Ti confirms the well-known fact that the structural/compositional component of resistance in alloyed systems is higher by one order of magnitude than the temperature-dependent component [69]. The decrease of the resistance in Ti-6Al-7Nb alloy above 700°C is caused by increasing equilibrium amount of β-phase with increasing temperature. Note that the Ti-6Al-4V alloy containing approximately 15% of β-phase particles at 750°C and 20% of β-phase particles at 800°C exhibited the similar resistance decrease [70]. The most important result is the obvious difference in resistance evolution between UFG materials and their coarse grained counterparts. This difference is more apparent in the Ti-6Al-7Nb alloy and is probably caused by more pronounced structural effect on overall resistance than in CP Ti. In the CP Ti, the difference in resistance evolution is almost certainly caused by recovery and/or recrystallization as no structure changes occur in the investigated temperature range. We assume, however, that recrystallization and/or recovery is also responsible for the differences in resistivity evolution in Ti-6Al-7Nb. However, in this alloy, other effects including changes in β-phase particles morphology, reduced amount of phase interfaces, and also increasing equilibrium amount of β-phase at elevated temperatures are expected to affect the overall resistance.

4.2.2. Correlation between mechanical properties and microstructure

The CP Ti processed by ECAP exhibits nearly constant value of microhardness after annealing at temperatures lower than 450–500°**C** (**Figure 3**). In this temperature range, the recovery of the material starts and the microhardness declines. The microhardness data are consistent with the electrical resistance measurements. Observations by scanning electron microscopy (**Figure 6**) revealed that material recovery/recrystallization is responsible for the first bump in the first derivative of resistance and the decrease of materials microhardness.

Similarly to the CP Ti, the microhardness of the UFG Ti-6Al-7Nb alloy remains constant during heating up to 440 and 550°**C**. Note that both annealed and UFG samples of Ti-6Al-7Nb alloy were heat treated at 500°**C** for 1 hour, which is considered as a strength increasing heat treatment [71, 72]. Heating of UFG samples up to 440 and 550°**C** does not affect the microhardness, despite an obvious response of the electrical resistivity, which can be probably attributed to the recovery process. It is therefore assumed that an initial stage of the recovery process has only a negligible effect on the microhardness in Ti-6Al-7Nb alloy. The annealing of the sample up to 660°**C** leads to a slight decrease of the microhardness. The effect of heating on microhardness is much lower in Ti-6Al-7Nb alloy than in CP Ti due to solid solution strengthening and, even more importantly, due to strengthening by phase interfaces.

SEM observations of CP Ti did not reveal any microstructural changes after heating up to 440°**C**. It is consistent with the electrical resistance evolution and microhardness measurements and also with the results of other authors [73]. Thermally activated processes in CP Ti during annealing up to 440°**C** were not observed. Further annealing to 520°**C** caused significant recovery and possibly even the initial stage of recrystallization/grain growth. These processes are responsible for significant decrease of microhardness. Annealing up to 640°**C** caused complete recovery and recrystallization. Such processes were apparently detected by *in-situ* measurement of electrical resistance. The results proved that high sensitivity *in-situ* measurement of electrical resistance is capable of detecting recovery and/or recrystallization processes in temperature regions that are decisive for microstructure stability of UFG CP Ti.

The comparison of resistance measurements and SEM observations is less convincing in Ti-6Al-7Nb alloy than in CP Ti. The microstructure remains unchanged after annealing up to 440°C, which is consistent with the resistance measurements. Despite resistance evolution suggests a microstructural transformation in the condition annealed to 550°C, no obvious microstructure changes were observed. On the other hand, other authors reported recovery process (identified by X-ray diffraction) and even the beginning of recrystallization (observed by TEM) in Ti-6Al-7Nb alloy prepared by ECAP and annealed at 500°C for 1 h [33]. However, our sample heated up to 550°C at a constant rate of 5°C min^{-1} was in fact exposed to temperatures above 500°C only for 10 min. This relatively short time of exposure to temperatures above 500°C might be insufficient for recovery process to be observed by SEM. Partially recrystallized structure of the sample annealed up to 660°C is shown in **Figure 6(d)** and is consistent with the results in Ref. [33], in which resistance measurements and observations of the similar UFG material annealed at 600°C for 1 h are reported. Therefore, we are convinced that electrical resistance measurement captured the recovery and recrystallization processes also in Ti-6Al-7Nb, despite the beginning of the process could not be unambiguously proven by SEM observations.

5. Conclusion

Evolution of microstructure of ultra-fine grained magnesium alloy AZ31, CP Ti (Grade 4), and Ti-6Al-7Nb alloy prepared by equal-channel angular pressing was investigated. Several experimental techniques were employed in order to identify the processes operating during heating of the material. The following conclusions can be drawn from this study:

- Ultra-fine grained titanium and magnesium-based materials were successfully prepared by ECAP. Microstructural refinement significantly increases their microhardness.
- UFG AZ31 alloy is stable up to 170°C (0.5 T_m), while CP Ti and Ti-6Al-7Nb alloy are stable approximately to 500°C, which corresponds to 0.4 T_m.
- The decrease of the microhardness upon annealing of both AZ31 alloy and CP Ti was attributed to annihilation of dislocation and subsequent grain growth.
- Recovery processes in Ti-6Al-7Nb alloy occur in the similar temperature range as in CP Ti. However, Ti-6Al-7Nb does not exhibit strong decrease of microhardness even after heating to 660°C, which is associated with solid solution strengthening of the material by phase interfaces.
- *In-situ* electrical resistance measurement is capable to detect recovery and/or recrystallization processes. It revealed differences in resistance evolution between ultra-fine grained and coarse grained condition of Ti-based materials. These differences correspond to microstructural changes observed by SEM and associated microhardness decrease.
- The kinetics of grain growth of AZ31 alloy was described by the Arrhenius equation and the activation energies of grain growth were determined. The values of activation energy continuously increase with increasing temperature due to changes of dominant diffusion mechanisms.

Acknowledgements

This work was financially supported by ERDF under the project "Nanomaterials centre for advanced applications," project No. CZ.02.1.01/0.0/0.0/15_003/0000485. J. Strásk‌á also acknowledges Czech Science Foundation under the project 16-08963S. J. Stráský, K. Václavová, and P. Zháňal acknowledge Czech Science Foundation under the project 17-20700Y.

Author details

Jitka Strásk‌á[1]*, Pavel Zháňal[1], Kristína Václavová[1], Josef Stráský[1], Petr Harcuba[1], Jakub Čížek[2] and Miloš Janeček[1]

*Address all correspondence to: straska.jitka@gmail.com

1 Department of Physics of Materials, Charles University, Prague, Czech Republic

2 Department of Low-Temperature Physics, Charles University, Prague, Czech Republic

References

[1] Gupta M, Sharon NML. In: Magnesium, Magnesium Alloys, and Magnesium Composites. New Jersey: John Wiley & Sons, Inc.; 2011. pp. 1-3

[2] Geetha M, Singh AK, Askomanani R, Gugia AK. Ti based biomaterials, the ultimate choice for orthopaedic implants—Review. Progress in Materials Science. 2009;**54**:397-425. DOI: 10.1016/j.pmatsci.2008.06.004

[3] Long M, Rack HJ. Titanium alloys in total joint replacement—A materials science perspective. Biomaterials. 1998;**19**:1621. DOI: 10.1016/S0142-9612(97)00146-4

[4] Elias CN, Meyers MA, Valiev RZ, Monteiro SN. Ultrafine grained titanium for biomedical applications: An overview of performance. Journal of Materials Research and Technology. 2013;**2**:340-350. DOI: 10.1016/j.jmrt.2013.07.003

[5] Welsch G, Boyer R, Collings EW, editors. Materials Properties Handbook: Titanium Alloys. ASM International, Materials Park, Ohio 44073-0002, USA; 1993

[6] Eisenbarth E, Velten D, Müller M, Thull R, Breme J. Biocompatibility of β-stabilizing elements of titanium alloys. Biomaterials. 2004;**25**:5705-5713. DOI: 10.1016/j.biomaterials.2004.01.021

[7] Yu J, Zhao ZJ, Li LX. Corrosion fatigue resistances of surgical implant stainless steel and titanium alloys. Corrosion Science. 1993;**35**:587-591. DOI: 10.1016/0010-938X(93)90193-K

[8] Leyens C, Peters M. Titanium and Titanium Alloys. Weinheim: Wiley-VCH; 2003. p. 2. DOI: 10.1002/3527602119

[9] Janeček M, Stráský J, Čížek J, Harcuba P, Václavová K, Polyakova VV, Semenova IP. Mechanical properties and dislocation structure evolution in Ti6Al7Nb alloy processed by high pressure torsion. Metallurgical and Materials Transactions A. 2014;**45A**:7-15. DOI: 10.4028/www.scientific.net/MSF.667-669.943

[10] Polyakova VV, Anumalasetty VN, Semenova IP, Valiev RZ. Influence of UFG structure formation on mechanical and fatigue properties in Ti-6Al-7Nb alloy. IOP Conference Series: Materials Science and Engineering. 2014;**63**:1-8. DOI: 10.1088/1757-899X/63/1/012162

[11] Valiev RZ, Islamgaliev RK, Alexandrov IV. Bulk nanostructured materials from severe plastic deformation. Progress in Materials Science. 2000;**45**:103-189. DOI: 10.1016/S0079-6425(99)00007-9

[12] Langdon TG. Twenty-five years of ultrafine-grained materials: Achieving exceptional properties through grain refinement. Acta Materialia. 2013;**61**:7035-7059. DOI: 10.1016/j.actamat.2013.08.018

[13] Valiev RZ, Estrin Y, Horita Z, Langdon TG, Zehetbauer MJ, Zhu YT. Producing bulk ultrafine-grained materials by severe plastic deformation. Journal of Management. 2006;**58**:33-39. DOI: 10.1007/s11837-006-0213-7

[14] Valiev RZ, Langdon TG. Principles of equal-channel angular pressing as a processing tool for grain refinement. Progress in Materials Science. 2006;**51**:881-981. DOI: 10.1016/j.pmatsci.2006.02.003

[15] Matsubara K, Miyahara Y, Horita Z, Langdon TG. Developing superplasticity in a magnesium alloy through a combination of extrusion and ECAP. Acta Materialia. 2003;**51**:3073-3084. DOI: 10.1016/S13596454(03)00118-6

[16] Zhilyaev AP, Langdon TG. Using high-pressure torsion for metal processing: Fundamentals and applications. Progress in Materials Science. 2008;**53**:893-979. DOI: 10.1016/j.pmatsci.2008.03.002

[17] Saito Y, Utsunomiya H, Tsuji N, Sakai T. Novel ultra-high straining process for bulk materials—Development of the accumulative roll-bonding (ARB) process. Acta Materialia. 1999;**47**:579-583. DOI: 10.1016/S1359-6454(98)00365-6

[18] Horita Z, Fujinami T, Langdon TG. The potential for scaling ECAP: Effect of sample size on grain refinement and mechanical properties. Materials Science and Engineering: A. 2001;**318**:34-41. DOI: 10.1016/S0921-5093(01)01339-9

[19] Agnew SR, Horton JA, Lillo TM, Brown DW. Enhanced ductility in strongly textured magnesium produced by equal channel angular processing. Scripta Materialia. 2004;**50**:377-381. DOI: 10.1016/j.scriptamat.2003.10.006

[20] Janeček M, Popov M, Krieger MG, Hellmig RJ, Estrin Y. Mechanical properties and microstructure of a Mg alloy AZ31 prepared by equal-channel angular pressing. Materials Science and Engineering: A. 2007;**462**:116-120. DOI: 10.1016/j.msea.2006.01.174

[21] Figueiredo RB, Langdon TG. Developing superplasticity in a magnesium AZ31 alloy by ECAP. Journal of Materials Science. 2008;**43**:7366-7371. DOI: 10.1007/s1085-008-2846-0

[22] Figueiredo RB, Langdon TG. Grain refinement and mechanical behavior of a magnesium alloy processed by ECAP. Journal of Materials Science. 2010;**45**:4827-4836. DOI: 10.1007/s10853-010-4589-y

[23] Xu J, Shirooyeh M, Wongsa-Ngam J, Shan D, Guo B, Langdon TG. Hardness homogeneity and micro-tensile behavior in a magnesium AZ31 alloy processed by equal-channel angular pressing. Materials Science and Engineering: A. 2013;**586**:108-114. DOI: 10.1016/j.msea.2013.07.096

[24] Stolyarov VV, Zhu YT, Lowe TC, Islamgaliev RK, Valiev RZ. A two-step SPD processing of ultrafine-grained titanium. Nanostructured Materials. 1999;**11**:947-954. DOI: 10.1016/S0965-9773(99)00384-0

[25] Islamgaliev RK, Kazyhanov VI, Shestakova LI, Sharafurdinov AV, Valiev RZ. Microstructure and mechanical properties of titanium (Grade 4) processed by high-pressure torsion. Materials Science and Engineering: A. 2008;**493**:190-194. DOI: 10.1016/j.msea.2007.08.084

[26] Wang YC, Langdon TG. Influence of phase volume fractions on the processing of a Ti-6Al-4V alloy by high-pressure torsion. Materials Science and Engineering: A. 2013;**559**:861-867. DOI: 10.1016/j.msea.2012.09.034

[27] Gubicza J, Chinh NQ, Dobatkin SV, Khosravi E, Langdon TG. Stability of ultrafine-grained microstructure in fcc metals processed by severe plastic deformation. Key Engineering Materials 2011;**465**:195-198. DOI: 10.4028/www.scientific.net/KEM.465.195

[28] Kim HK. Activation energies for the grain growth of an AZ31 Mg alloy after equal channel angular pressing. Journal of Materials Science (Letters). 2004;**39**:7107-7109. DOI: 10.1023/B:JMSC.0000047560.93940.45

[29] Radi Y, Mahmudi R. Effect of Al_2O_3 nano-particles on the microstructural stability of AZ31 Mg alloy after equal channel angular pressing. Materials Science and Engineering: A. 2010;**527**:2764-2771. DOI: 10.1016/j.msea.2010.01.029

[30] Vrátná J, Janeček M, Čížek J, Lee DJ, Yoon EY, Kim HS. Mechanical properties and microstructure evolution in ultrafine-grained AZ31 alloy processed by severe plastic deformation. Journal of Materials Science. 2013;**48**:4705-4712. DOI: 10.1007/s10853-013-7151-x

[31] Zháňal P, Václavová K, Hadzima B, Harcuba P, Stráský J, Janeček M, Polyakova V, Semenova I, Hájek M, Hajizadeh K. Thermal stability of ultrafine-grained commercial purity Ti and Ti–6Al–7Nb alloy investigated by electrical resistance, microhardness and scanning electron microscopy. Materials Science and Engineering: A. 2016;**651**:886-892. DOI: 10.1016/j.msea.2015.11.029

[32] Hajizadeh K, Eghbali B, Topolski K, Kurzydlowski KJ. Ultra-fine grained bulk CP-Ti processed by multi-pass ECAP at warm deformation region. Materials Chemistry and Physics. 2014;**143**:1032-1038. DOI: 10.1016/j.matchemphys.2013.11.001

[33] Polyakova V, Semenova IP, Valiev RZ. Influence of annealing on the structure and mechanical properties of ultrafine-grained alloy Ti–6Al–7Nb, processed by severe plastic deformation. Materials Science Forum. 2010;**667-669**:943-948. DOI: 10.4028/www.scientific.net/MSF.667-669.943

[34] Hájek M, Veselý J, Cieslar M. Precision of electrical resistivity measurements. Materials Science and Engineering: A. 2007;**462**:339-342. DOI:10.1016/j.msea.2006.01.175

[35] Vrátná J, Janeček M. Experimental study of ultrafine-grained commercial magnesium alloy AZ31 prepared by severe plastic deformation. In: Proceeding of the 22st International Conference on Metallurgy and Material (METAL 2013). 2014. ISBN 978-80-87294-39-0

[36] Stráská J, Janeček M, Čížek J, Stráský J, Hadzima B. Microstructure stability of ultra-fine grained magnesium alloy AZ31 processed by extrusion and equal-channel angular pressing (EX–ECAP). Materials Characterization. 2014;**94**:69-79. DOI: 10.1016/j.matchar.2014.05.013

[37] Aghababaei R, Joshi SP. Micromechanics of tensile twinning in magnesium gleaned from molecular dynamics simulations. Acta Materialia. 2014;**69**:326-342. DOI: 10.1016/j.actamat.2014.01.014

[38] Kazemi-Choobi K, Khalil-Allafi J, Abbasi-Chianeh V. Investigation of the recovery and recrystallization processes of Ni50.9Ti49.1 shape memory wires using in situ electrical resistance measurement. Materials Science and Engineering: A. 2012;**551**:122-127. DOI: 10.1016/j.msea.2012.04.106

[39] Stráská J, Stráský J, Minárik P, Janeček M, Hadzima B. Continuous measurement of *m*-parameter for analyzing plastic instability in a superplastic ultra-fine grained magnesium alloy. Materials Science and Engineering: A. 2017;**684**:110-114. DOI: 10.1016/j.msea.2016.12.027

[40] Huang Y, Langdon TG. Characterization of deformation processes in a Zn-22% Al alloy using atomic force microscopy. Journal of Materials Science. 2002;**37**:4993-4998. DOI: 10.1023/A:1021071228521

[41] Vinogradov A, Hashimoto S, Patlan V, Kitagawa K. Atomic force microscopic study on surface morphology of ultra-fine grained materials after tensile testing. Materials Science and Engineering: A. 2001;**319-321**:862-866. DOI: 10.1016/S0921-5093(01)01057-7

[42] Han JH, Mohamed FA. Quantitative measurements of grain boundary sliding in an ultrafine-grained Al alloy by atomic force microscopy. Metallurgical and Materials Transactions A. 2011;**42**:3969-3978. DOI: 10.1007/s11661-011-0871-0

[43] Estrin Y, Tóth LS, Molinari A, Bréchet Y. A dislocation-based model for all hardening stages in large strain deformation. Acta Materialia. 1998;**46**:5509-5522. DOI: 10.1016/S1359-6454(98)00196-7

[44] Hall EO. The deformation and ageing of mild steel: III discussion of results. Proceedings of the Royal Society. 1951;**B64**:747-753. DOI: 10.1088/0370-1301/64/9/303

[45] Petch NJ. The cleavage strength of polycristals. The Journal of the Iron and Steel Institute. 1953;**174**:25-27

[46] Higgins GT. Grain-boundary migration and grain growth. Metal Science. 1974;**8**:143-150

[47] Kim HK, Kim WJ. Microstructural instability and strength of an AZ31 Mg alloy after severe plastic deformation. Materials Science and Engineering: A. 2004;**385**:300-308. DOI: 10.1016/j.msea.2004.06.055

[48] Cao P, Lu L, Lai MO. Grain growth and kinetics for nanocrystalline magnesium alloy produced by mechanical alloying. Materials Research Bulletin. 2001;**36**:981-988. DOI: 10.1016/S0025-5408(01)00578-5

[49] Thein MA, Lu L, Lai MO. Kinetics of grain growth in nanocrystalline magnesium-based metal–metal composite synthesized by mechanical alloying. Composites Science and Technology. 2006;**66**:531-537. DOI: 10.1016/j.compscitech.2005.07.002

[50] Miao Q, Hu L, Wang X, Wang E. Grain growth kinetics of a fine-grained AZ31 magnesium alloy produced by hot rolling. Journal of Alloys and Compounds. 2010;**493**:87-90. DOI: 10.1016/j.jallcom.2009.12.049

[51] Chao HY, Sun HF, Chen WZ, Wang ED. Static recrystallization kinetics of a heavily cold drawn AZ31 magnesium alloy under annealing treatment. Materials Characterization. 2011;**62**:312-320. DOI: 10.1016/j.matchar.2011.01.007

[52] Frost HJ, Ashby MF. Deformation-Mechanism Maps. Oxford: Pergamon Press; 1982. pp. 43-45

[53] Shewman PG. Self-diffusion in Mg single crystals. Trans AIME. 1956;**206**:918-22

[54] Wang J, Iwahashi Y, Horita Z, Furukawa M, Nemoto M, Valiev RZ, Langdon TG. An investigation of microstructural stability in an AlMg alloy with submicrometer grain size. Acta Materialia. 1996;**44**:2973-2982. DOI: 10.1016/1359-6454(95)00395-9

[55] Valiev RZ, Kozlov EV, Ivanov YF, Lian J, Nazarov AA, Baudelet B. Deformation behaviour of ultra-fine-grained copper. Acta Metallurgica et Materialia. 1994;**42**:2467-2475. DOI: 10.1016/0956-7151(94)90326-3

[56] Lian J, Valiev RZ, Baudelet B. On the enhanced grain growth in ultrafine grained metals. Acta Metallurgica et Materialia. 1995;**43**:4165-4170. DOI: 10.1016/0956-7151(95)00087-C

[57] Janeček M, Yi S, Král R, Vrátná J, Kainer KU. Texture and microstructure evolution in ultrafine-grained AZ31 processed by EX-ECAP. Journal of Materials Science. 2010;**45**:4665-4671. DOI: 10.1007/s10853-010-4675-1

[58] Stráská J, Stráský J, Janeček M. Activation energy for grain growth of the isochronally annealed ultrafine grained magnesium alloy after hot extrusion and equal-channel angular pressing (EX-ECAP). Acta Physica Polonica A. 2015;**128**:578-581

[59] Furukawa M, Horita Z, Nemoto M, Valiev RZ. Microhardness measurements and the Hall-Petch relationship in an Al+Mg alloy with submicrometer grain size. Acta Materialia. 1996;**51**:4619-4629. DOI: 10.1016/1359-6454(96)00105-X

[60] Vrátná J, Janeček M, Stráský J, Kim HS, Yoon EY. Investigation of microhardness and microstructure of AZ31 alloy after high pressure torsion. Magnesium Technology, Magnesium Technology Series. 2011:589-594

[61] Lin HK, Huang JC. High strain rate and/or low temperature superplasticity in AZ31 Mg alloys processed by simple high-ratio extrusion methods. Materials Transactions. 2002;**43**:2424-2432

[62] Lin HK, Huang JC, Langdon TG. Relationship between texture and low temperature superplasticity in an extruded AZ31 Mg alloy processed by ECAP. Materials Science and Engineering: A. 2005;**402**:250-257. DOI: 10.1016/j.msea.2005.04.018

[63] Figueiredo RB, Langdon TG. Principles of grain refinement and superplastic flow in magnesium alloys processed by ECAP. Materials Science and Engineering: A. 2009;**501**:105-114. DOI: 10.1016/j.msea.2008.09.058

[64] Mohan A, Yuan W, Mishra RS. High strain rate superplasticity in friction stir processed ultrafine grained Mg-Al-Zn alloys. Materials Science and Engineering: A. 2013;**562**:69-76. DOI: 10.1016/j.msea.2012.11.026

[65] Janeček M, Čížek J, Gubicza J, Vrátná J. Microstructure and dislocation density evolutions in MgAlZn alloy processed by severe plastic deformation. Journal of Materials Science. 2012;**47**:7860-7869. DOI: 10.1007/s10853-012-6538-4

[66] Young JP, Askari H, Hovanski Y, Heiden MJ, Field DP. Thermal microstructural stability of AZ31 magnesium after severe plastic deformation. Materials Characterization. 2015;**101**:9-19. DOI: 10.1016/j.matchar.2014.12.026

[67] Langdon TG. The mechanical properties of superplastic materials. Metallurgical and Materials Transactions A. 1982;**13**:689-701. DOI: 10.1007/BF02642383

[68] Cormier M, Claisse F. Composition dependence of the thermal coefficient of electrical resistivity of c.p.h. titanium-oxygen alloys. Journal of the Less Common Metals. 1976;**48**:309-314. DOI: 10.1016/0022-5088(76)90010-2

[69] Powell RW, Tye RP. The thermal and electrical conductivity of titanium and its alloys. Journal of the Less Common Metals. 1961;**48**:309-314. DOI: 10.1016/0022-5088(61)90064-9

[70] Malinov S, Markovsky P, Sha W, Guo Z. Resistivity study and computer modelling of the isothermal transformation kinetics of Ti-6Al-4V and Ti-6Al-2Sn-4Zr-2Mo-0.08Si alloys. Journal of Alloys and Compounds. 2001;**314**:181-192. DOI: 10.1016/S0925-8388(00)01227-5

[71] Welsch G, Lütjering G, Gazioglu K, Bunk W. Deformation characteristics of age hardened Ti-6Al-4V. Metallurgical and Materials Transactions A. 1977;**8**:169-177. DOI: 10.1007/BF02677278

[72] Guo S, Meng Q, Liao G, Zhao X. Microstructural evolution and mechanical behaviour of metastable β-type Ti-25Nb-2Mo-4Sn alloy with high strength and low modulus. Progress in Natural Science: Materials. 2013;**23**:174-182. DOI: 10.1016/j.pnsc.2013.03.008

[73] Hoseini M, Pourian MH, Bridier F, Vali H, Szpunar JA, Bocher P. Thermal stability and annealing behaviour of ultrafine commercially pure titanium. Materials Science and Engineering: A. 2012;**532**:58-63. DOI: 10.1016/j.msea.2011.10.062

Chapter 7

New Combined Technology of Deformation "Rolling-Equal Channel Angular Pressing", Allowing to Obtain Metals and Alloys with Sub-Ultra-fine-Grained Structure

Abdrakhman Naizabekov, Sergey Lezhnev,
Evgeniy Panin and Irina Volokitina

Additional information is available at the end of the chapter

http://dx.doi.org/10.5772/intechopen.68663

Abstract

In this chapter, results of theoretical and laboratory research of combined process "rolling-pressing" using equal-channel step matrix are described. In theoretical studies, the empirical dependences for determining forces of rolling, extrusion, and back pressure in the matrix have been obtained. A program was compiled for finding the optimal value of the angle of intersection of channels in the matrix. In the study of kinematic parameters of the process were obtained formulas for determining diameters of the rolls, the values of which would provide the best capture angle of the workpiece. During computer simulation, the parameters of stress-strain state, temperature distribution, and the influence of the tilting of the workpiece on the process parameters were studied. In the laboratory experiment, the effect of the new combined deformation process "rolling-pressing" on the evolution of microstructure of copper was studied.

Keywords: combined process, rolling-pressing, modeling, experiment, microstructure, copper

1. Introduction

The research and development studies aimed at obtaining high-strength metals and alloys are currently of great scientific and practical interest. Production of metals with unique properties may be possible by reduction of grain through the implementation of intensive plastic deformation in the whole bulk of the deformable workpiece.

A range of new processes of metal-forming designed to produce metal with sub-ultrafine-grained structure whose basic principle refers to the realization of a simple shear scheme in the process of deformation has been recently developed. The equal channel angular pressing of billets [1] in the matrices of various designs is one of these methods. It provides an intensive plastic deformation without significant change of the original cross-sectional dimensions of the workpiece. However, it has a significant drawback—the length of the original billet is limited by the space of the equipment used, i.e., by the working stroke of the press punch. Furthermore, this method of deformation does not provide a continuous pressing process. That is why this method of obtaining a sub-ultrafine-grained metal structure has not found an industrial implementation. It is still in an object of investigation using relatively small samples under laboratory conditions.

The aim of this work is to develop and investigate a new method of deformation, allowing production of long-length workpieces with sub-ultrafine-grained structure at low energy costs.

2. Theory, materials, and research methods

For this purpose, at "Metal forming" Department of Karaganda State Industrial University, the combined process "rolling-pressing" with the use of equal-channel step matrix, calibrated [2] and smooth [3] rolls (**Figure 1**), was developed, which in comparison with conventional compression in equal-channel step matrix partially removes limitations on the sizes of initial billets.

The essence of the proposed method of deformation is as follows. The workpiece which is preheated at the beginning of deformation is applied to forming roll and is fed to the rolling rolls through the contact friction forces capturing it in roll gap. At output, it is pushed through the channels of an equal-channel step matrix. The next billet is served at the time when the workpiece is fully released from the gap of the rolls. Passing through the rolls and once in the matrix, it pushes out of the matrix the billet previously deformed. The process of pressing in this case is implemented through the use of contact friction forces acting at the surface of contact between the metal and the rotating rolls. This process, as noted above, can be applied using smooth and calibrated rolls. The comparative analysis shows that the later use is the

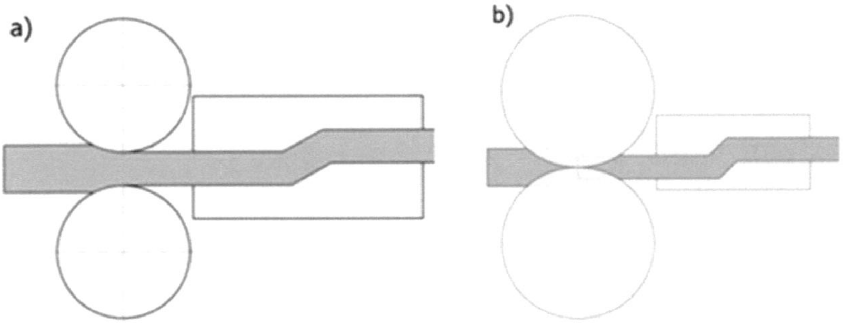

Figure 1. Scheme of combined process "rolling-pressing" with smooth rolls (a) and calibrated rolls (b).

most optimal solution, because it requires a smaller absolute compression of the billet during its rolling to push it through the channels of the matrix. Besides, the broadening of the workpiece in the rolls can be controlled, which in the case of using of smooth rolls should be calculated [4].

The feasibility of the "rolling-pressing" process requires that the projection on the X-axis of the sum of forces generated by the rolls in the deformation zone (marked as P_{ROLL}) is greater than the effort required to push the billet through the channels of the matrix (marked as P_{PRESS}), i.e.:

$$P_{ROLL} > P_{PRESS} \tag{1}$$

The projection on the X-axis of the sum of all forces acting at the deformation zone (**Figure 2**) is described by

$$P_{ROLL} = 2b_{av}\int_{\gamma_1}^{\alpha}\tau_{av}R\cos\theta d\theta - 2b_{av}\int_{0}^{\gamma}\tau_{av}R\cos\theta d\theta - 2b_{av}\int_{0}^{\alpha}p_{av}R\sin\theta d\theta \tag{2}$$

where b_1 and b_{av} are the width of the workpiece after deformation and average width; τ_{av} and p_{av} are the average tangential and normal stresses; R is the radius of the rolls; θ is the current angle; α is the angle of capture; γ and γ_1 are the angles characterizing the extent of zones of advance and lag, respectively.

Angles γ and γ_1 can be found using the methodology described in the work [5]. Integrate Eq. (2) assuming that $b_{av} = b_1$:

$$P_{ROLL} = 2b_1R\tau_{AV}(\sin\alpha - \sin\gamma_1) - 2b_1R\tau_{AV}(\sin\gamma - 0) - 2b_1Rp_{AV}(-\cos\alpha + 1) \tag{3}$$

Replacing in this equation $1 - \cos\alpha = \frac{\alpha^2}{2}$, $\sin\alpha = \alpha$, $\sin\gamma_1 = \gamma_1$, $\sin\gamma = \gamma$, and taking into account that $\tau_{AV} = p_{AV}\mu = \sigma_S\mu$, we get

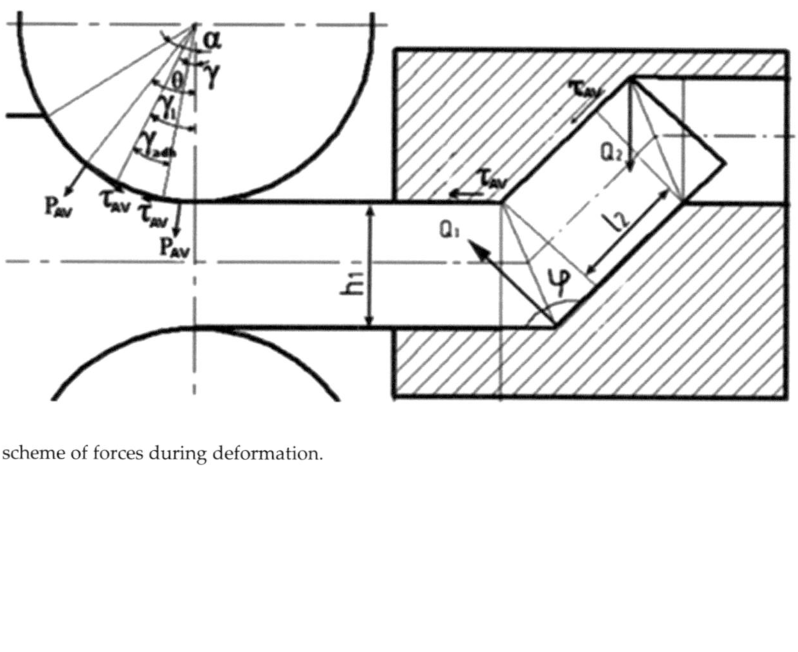

Figure 2. The scheme of forces during deformation.

$$P_{\mathrm{ROLL}} = 2b_1 R\sigma_S \mu_1 (\alpha - \gamma_1) - 2b_1 R\sigma_S \mu_1 \gamma - 2b_1 R\sigma_S \frac{\alpha^2}{2} \tag{4}$$

where μ_1 is the coefficient of friction in the deformation zone during rolling. Finally, we obtain

$$P_{\mathrm{ROLL}} = 2b_1 R\sigma_S \mu_1 \left(\alpha - \gamma_1 - \gamma - \frac{\alpha^2}{2\mu_1} \right) \tag{5}$$

When using grooved rolls, Eq. (5) takes the form:

$$P_{\mathrm{ROLL\_CAL}} = 2R\sigma_S \mu_1 \left[b_1 \left(\alpha - \gamma_1 - \gamma - \frac{\alpha^2}{2\mu_1} \right) + h_{\mathrm{av}} \alpha \right] \tag{6}$$

where h_{av} is the average height of workpiece in the deformation zone. In work [6], the formula for determining pressing forces in equal-channel angular matrix was obtained. Converting this formula taking into account the configuration of the channels of equal-channel step matrix, we obtain the formula for determining tonnage in this matrix:

$$P_{\mathrm{PRESS}} = 2\sigma_S \mu_2 \left[(2l_1 + l_2)(b_1 + h_1) + 2h_1^2 tg \frac{\phi}{2} + \frac{tg\phi \cdot h_1}{\sqrt{3}\mu_2} \right] \tag{7}$$

In the study of back pressure from the matrix, the following dependencies were obtained:

$$Q_1 = \left(2b_1 R\sigma_S \mu_1 \left(\alpha - \gamma_1 - \gamma - \frac{\alpha^2}{2\mu_1} \right) - \sigma_S \mu_2 \left[\left(2l_1 + h_1 ctg \frac{\phi}{2} \right) (b_1 + h_1) \right] \right) / \sin\phi \tag{8}$$

$$Q_2 = Q_1 \cos(180 - \phi) - 2\sigma_S \mu_2 \sin\phi \left(l_2 + h_1 ctg \frac{\phi}{2} \right) (b_1 + h_1) \tag{9}$$

Dependencies (8) and (9) fair when using rolls with smooth barrel. When using grooved rolls, it must also take account of forces of friction in the deformation zone on the side contact metal with the rolls. Taking into account the assumptions that were adopted when defining the force of backpressure Q_1 at the first stage of deformation with smooth rolls, force of backpressure Q_{1C} for process with calibrated rolls is determined by the formula:

$$Q_{1C} = \left(2R\sigma_S \mu_1 \left[b_1 \left(\alpha - \gamma_1 - \gamma - \frac{\alpha^2}{2\mu_1} \right) + h_{AV} \alpha \right] - \sigma_S \mu_2 \left[\left(2l_1 + h_1 ctg \frac{\phi}{2} \right) (b_1 + h_1) \right] \right) / \sin\phi \tag{10}$$

When projecting forces are on the vertical axis, all forces in the deformation zone cancel out. Therefore, Eq. (9) for determining the force of backpressure Q_{2C} in the case of grooved rolls is not changed. For the normal course of the process must ensure:

$$\frac{P_{\mathrm{PRESS}}}{h_1 b_1} < \sigma_S \tag{11}$$

Failure to comply with this condition, the metal in the area from the line connecting the centers of the rolls to equal-channel step matrix is repressed, increasing its transverse dimensions, thereby making the pressing process impossible.

One of the main factors influencing the pressing force is the angle of the junction of the channels of the matrix. To determine the optimal angle of junction of the channels of equal-channel step matrix, allowing the value of the compression force less than the force of rolling, a program in the Excel editor was compiled. This program allows us to determine the optimal angle of junction of graphically based plotting compression force depending on the angle of intersection of channels, and the rolling forces created by the rolls depending on the size of absolute compression. To comply with the condition (11), efforts were transferred to stresses:

$$\sigma_{\mathrm{ROLL}} = \frac{P_{\mathrm{ROLL}}}{h_1 b_1} \tag{12}$$

$$\sigma_{\mathrm{PRESS}} = \frac{P_{\mathrm{PRESS}}}{h_1 b_1} \tag{13}$$

where σ_{ROLL} is the stress in the cross section of the workpiece on the line connecting the centers of the rolls; σ_{PRESS} is the stress in the transverse section of the billet at the entrance to the matrix. As a result, graphs of σ_{ROLL}, σ_{PRESS}, and σ_S depending on the angle ϕ (**Figure 3**) were obtained.

Also, a study of kinematic parameters of this combined process was conducted. As noted above, a combined method of deformation of blanks "rolling-pressing" has advantages over previously known methods of pressing, but it has one drawback that it still does not ensure the

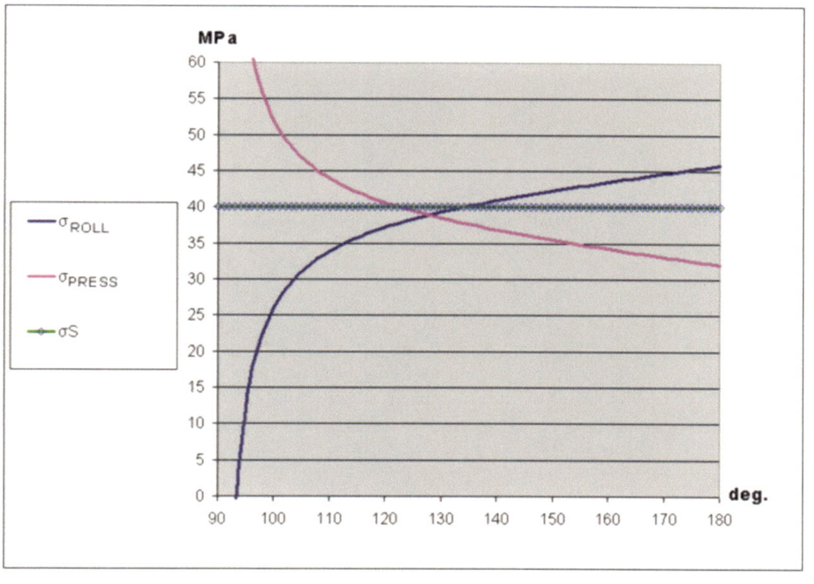

Figure 3. Graphs to determine the optimal angle of junction of the channels in the matrix.

continuity of the process, i.e., during deformation of a batch of blanks subsequent workpiece will push the previous. But after all deformation cycles, in the matrix will be the last not fully deformed workpiece. To remove this drawback, the scheme of combined process "rolling-pressing" with two pairs of rolls and equal-channel step matrix was proposed (**Figure 4**) [7, 8].

The essence of the proposed method of deformation is as follows. Preheated to a temperature of the beginning of the deformation the workpiece is fed to rolling rolls that captured it in roll gap by contact friction forces, and at the output from it, workpiece is pushed through the channels of equal-channel step matrix. After the workpiece exit from the channel matrix, it is captured by the second pair of rolls, which pull workpiece completely from the channels of the matrix. The advantage of this method is that during the implementation of this combined process, the proposed scheme ensures the continuity of the process and removes limitations on the sizes of initial billets.

After exiting from the matrix, the deformed billet will be redirected to the second pair of rolls that have to pull it out of the matrix; for capture of the workpiece by second pair of rolls, it needs to provide two conditions [9] as follows:

1. Optimal capture angle;
2. Optimal angular velocity of the rolls must be maintained so the workpiece that is in contact with the rolls is not jammed and slipped.

The final workpiece thickness (at the exit from the matrix) and roll diameters must be known to ensure an optimal capture. Determining the optimum speed of rotation of the rolls is a little complicated, since it is necessary to consider the influence of the matrix on the velocity of the workpiece. **Figure 5** shows the kinematic diagram of this method of deformation. Here, V_{01} is

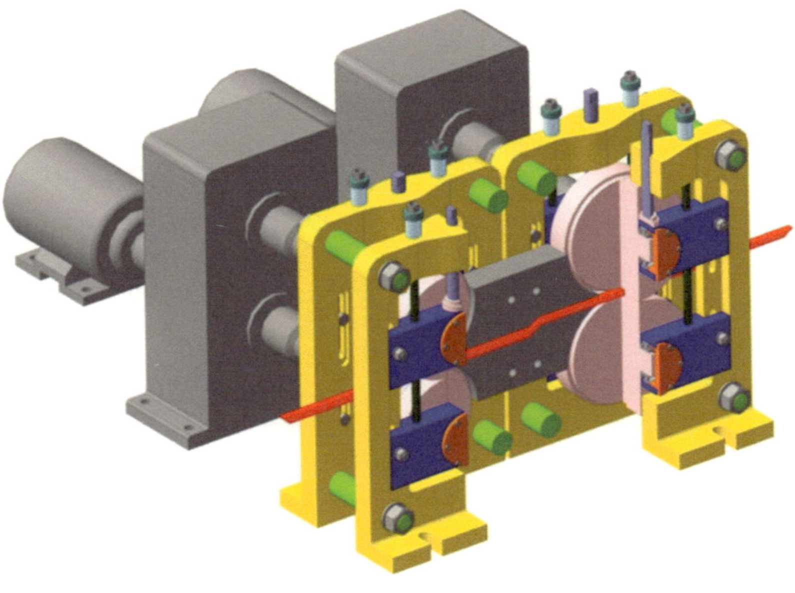

Figure 4. Combined process "rolling-pressing" with two pairs of rolls.

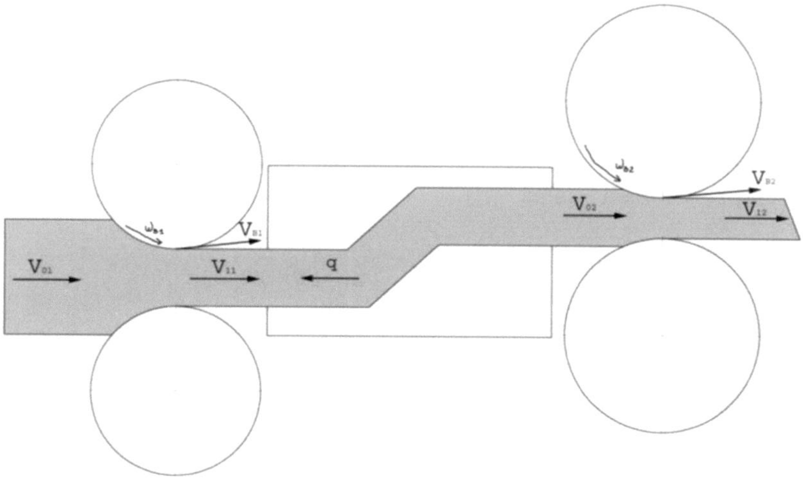

Figure 5. Kinematic diagram of the process "rolling-pressing" with two pairs of rolls.

the velocity of the metal at the entrance to the first pair of rolls; V_{11} is the velocity of the metal at the exit of the first pair of rolls; V_{02} is the velocity of the metal at the entrance of the second pair of rolls; V_{12} is the velocity of the metal at the exit of the second pair of rolls; $V_{\mathrm{R}1}$ is the rolling speed in the first pair of rolls; $V_{\mathrm{R}2}$ is the rolling speed in the second pair of rolls; $\omega_{\mathrm{R}1}$ is the circumferential speed of the first pair of rolls; $\omega_{\mathrm{R}2}$ is the circumferential speed of the second pair of rolls; q is the back pressure of matrix.

Equations were adopted on the following assumptions:

1. an optimum angle of capture was adopted by 18°;
2. as the whole construction is on one mill, we can take that both pairs of rolls are fed from the same electric motor. Therefore, the circumferential velocities of them are equal.

The result is the following formulas for finding the required diameter of the rolls:

$$D_2 = \frac{D_1(1 + S_1)}{\cos \alpha_2} \quad (14)$$

$$D_{2\mathrm{CAL}} = \frac{R_{\mathrm{ROL\_1}}(1 + S_{AV1})}{\cos \alpha_2} + h_1 \quad (15)$$

3. Computer simulation

The next stage of studying of this process was the simulation in program complex DEFORM. The purpose of this simulation was to investigate the stress-strain state of metal during the realization of combined process "rolling-pressing." Initially, the goal was to obtain a successful model in which the first pair of rolls has captured the workpiece, after which it crossed all channels of the matrix, at the output of which fell into the second pair of rolls which drew a blank from the matrix (**Figure 6**).

For investigation of the strain state, the study of parameter "strain effective" (**Figure 7**) was conducted, which includes the components of deformation in the following form:

$$\varepsilon_{EQ} = \frac{\sqrt{2}}{3}\sqrt{(\varepsilon_1 - \varepsilon_2)^2 + (\varepsilon_2 - \varepsilon_3)^2 + (\varepsilon_3 - \varepsilon_1)^2} \tag{16}$$

where ε_1, ε_2, and ε_3 are main strains.

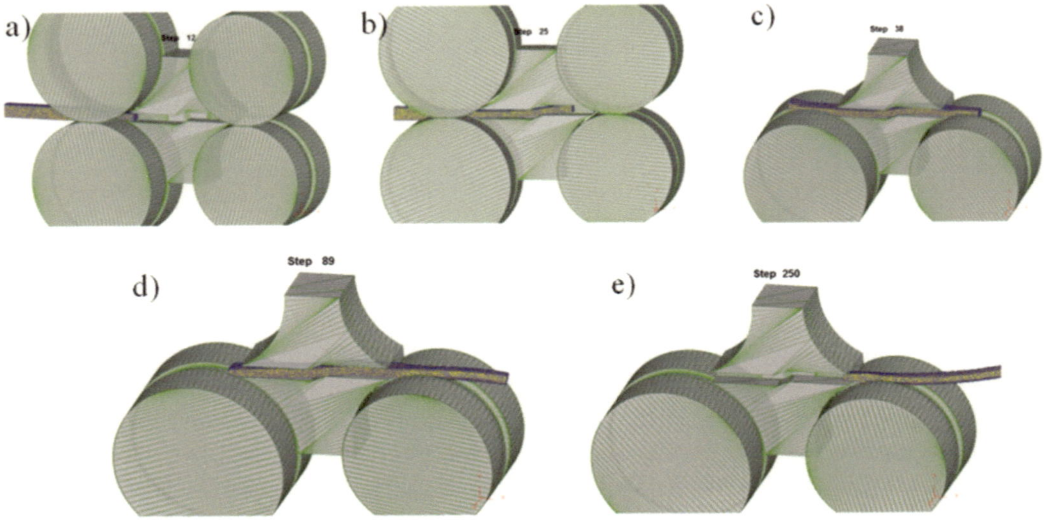

Figure 6. Stage of successful model with calibrated rolls: (a) the workpiece is in the 1st pair of rolls, (b) the workpiece is in the 1st pair of rolls and in the matrix, (c) the workpiece is in the both pairs of rolls and in the matrix, (d) the workpiece is in the matrix and in the 2nd pair of rolls, (e) the workpiece is in the 2nd pair of rolls.

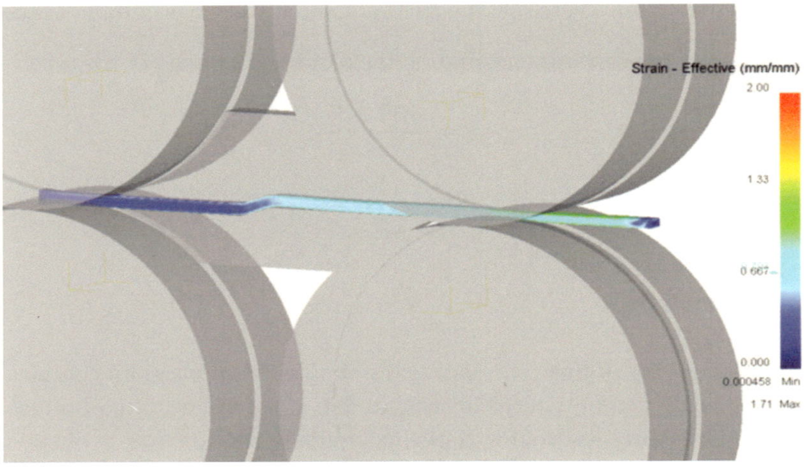

Figure 7. Strain state of the workpiece.

http://dx.doi.org/10.5772/intechopen.68663

For investigation of the stress state, the study of parameter "stress effective" (**Figure 8**) was conducted, which is defined as follows:

$$\sigma_{EQ} = \frac{1}{\sqrt{2}}\sqrt{(\sigma_1 - \sigma_2)^2 + (\sigma_2 - \sigma_3)^2 + (\sigma_3 - \sigma_1)^2} \tag{17}$$

where σ_1, σ_2, and σ_3 are main stresses.

For a detailed study of the stress and strain state, the modeling of process "rolling-pressing" was conducted by varying the main geometrical and technological factors that have a significant impact on the process. Analysis of the influence of various factors on the stress and strain state of this process showed that factors such as the angle of intersection of channels of the matrix, the coefficient of friction, temperature, and length of the channels of the matrix have a significant influence on the distribution of stresses and accumulated plastic deformation in the whole volume of the workpiece in the implementation of combined process.

The study of temperature conditions of this process (**Figure 9**) revealed that the temperature distribution over the cross section of the workpiece is uneven. A large temperature difference (up to 40°) can result in heterogeneity in physical properties. Therefore, to equalize the temperature difference in the cross section, it is recommended to carry out preheating of the matrix.

Figure 8. Stress state of the workpiece.

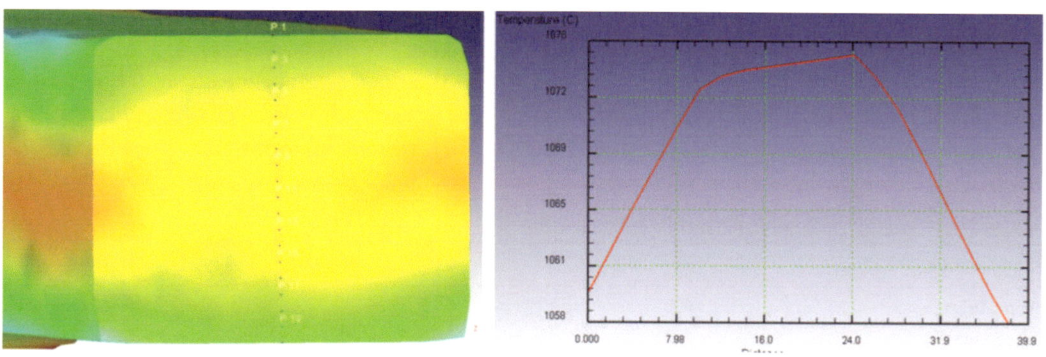

Figure 9. Study of temperature conditions.

In the study of the influence of the tilting of the workpiece (**Figure 10**) on the stress-strain state of the metal, it is established that the implementation of the tilting of the sample is not only has a positive effect on the distribution of accumulated plastic deformation and equivalent stresses in the workpiece, but also contributes to the restoration of the original shape of the cross section, which in some cases can play an important role.

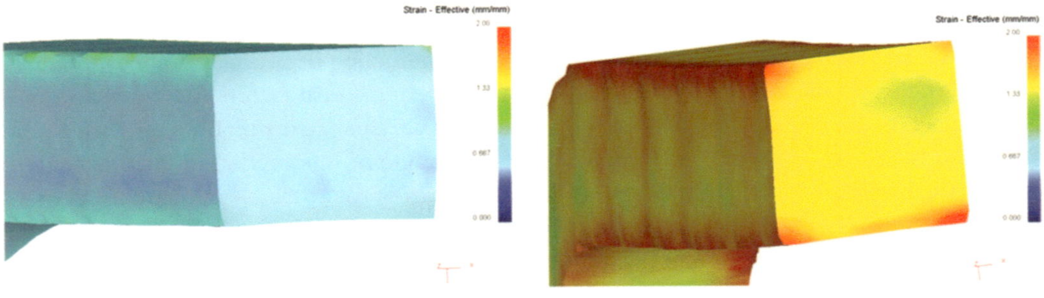

Figure 10. Impact of tilting on stress-strain state.

4. Laboratory experiments

After the simulation, laboratory experiments on the deformation of copper billets were carried by the combined process "rolling-pressing."" As material for the study, a copper alloy of M1 grade was taken. Before the "rolling-pressing" process, all samples were annealed, normalized, and tempered. Modes of heat treatment are given in **Table 1**. For the experiment, workpieces were made with dimensions $h \times b \times l = 16 \times 30 \times 200$ mm.

A laboratory experiment was conducted on designed stand for realization of the combined process "rolling-pressing" with the use of smooth rolls (**Figure 11**) based on rolling mill DUO-250. It was done in three passes.

At increasing temperatures, it is possible to start the process of grain growth during deformation. To exclude this growth, it is necessary to use the deformation temperature that is lower than the initial temperature of recrystallization [10]. Based on this fact, the experiment was conducted at ambient temperature.

With an increasing number of passes, the intensity of the dispersion of the structure increases, but it also increases the hardening of the material. As a result, the resource of plasticity is lost and further deformation and the use in industry of such a metal is impossible, since there is its destruction. For the reduction of the density of excess dislocations and increase of plasticity

Heat treatment mode	Temperature, °C	Time, min	Colling area
Annealing	600	16	Inside the oven
Normalizing	600	16	Air
Quenching	700	16	Water

Table 1. Heat treatment modes.

Figure 11. Experimental stand for realization of combined process "rolling-pressing."

resource, the metal should be heated at temperature that is lower than initial temperature of recrystallization. After determination of this temperature using equations [11], a laboratory experiment was performed. All billets after the "rolling-pressing" process were cut into pieces with a thickness of 5 mm and heated at temperatures of 100–270°C with duration of exposure 30 min and cooling in water.

The treated samples were studied in two sections: transverse and longitudinal, using optical and transmission electron microscopes (TEM). The resulting samples were also tested on a torsion-tensile machine to the test tension and compression. All the samples were studied in the mid-plane of the sample to avoid the influence of peripheral areas.

Preparation of thin sections for metallographic studies was carried out according to standard methods; for the study, an optical microscope Leica equipped with a set box was used for hardness testing. For the studies, at a transmission electron microscope JEOL JEM 2100, thin foils were prepared. For this purpose, from the sample using a precision cutting machine cutoff, the workpieces with a thickness of 250 μm that were subjected to fine grinding for removing the work-hardened layer. Then the samples were subjected to processing in the machine for electrolytic thinning Tenupol 5.

For determination the mechanical properties of copper after heat treatment and subsequent "rolling-pressing" process the torsion-testing machine MI40KU was used. For the testing, standard samples of cylindrical shape in quantities of 72 pieces were used, the diameter of the working part was 3 mm, working length was 15 mm, and tensile speed was 0.5 mm/min. This value corresponds to a strain rate equal to $0.56 \times 10^{-3}\ s^{-1}$.

5. Results and discussion

5.1. Microstructure of copper before and after the "rolling-pressing" process

Figure 12 presents pictures of the microstructure of copper in the initial state and after preliminary heat treatments. In the structure of the deformed copper, twins are clearly shown (**Figure 12a**), after deformation and annealing grains of copper have more equiaxed grain form (**Figure 12b**).

To evaluate the effectiveness of the "rolling-pressing" process, it is necessary to compare the microstructure of copper samples before and after deformation. Photographs of the microstructure, obtained in the study of copper after the "rolling-pressing" process, are presented in **Figure 13**.

Microstructural studies of deformed copper billets by the combined process showed that before deformation copper has a coarse structure with twins. Grains have an average size equal to 100 μm (**Figure 13**). After the first pass, the structure is intensely reduced up to 40 μm compared with its initial state. In the transverse cross section, the structure is homogeneous and has equiaxed grains. However, in the longitudinal section, the structure has strokefest in

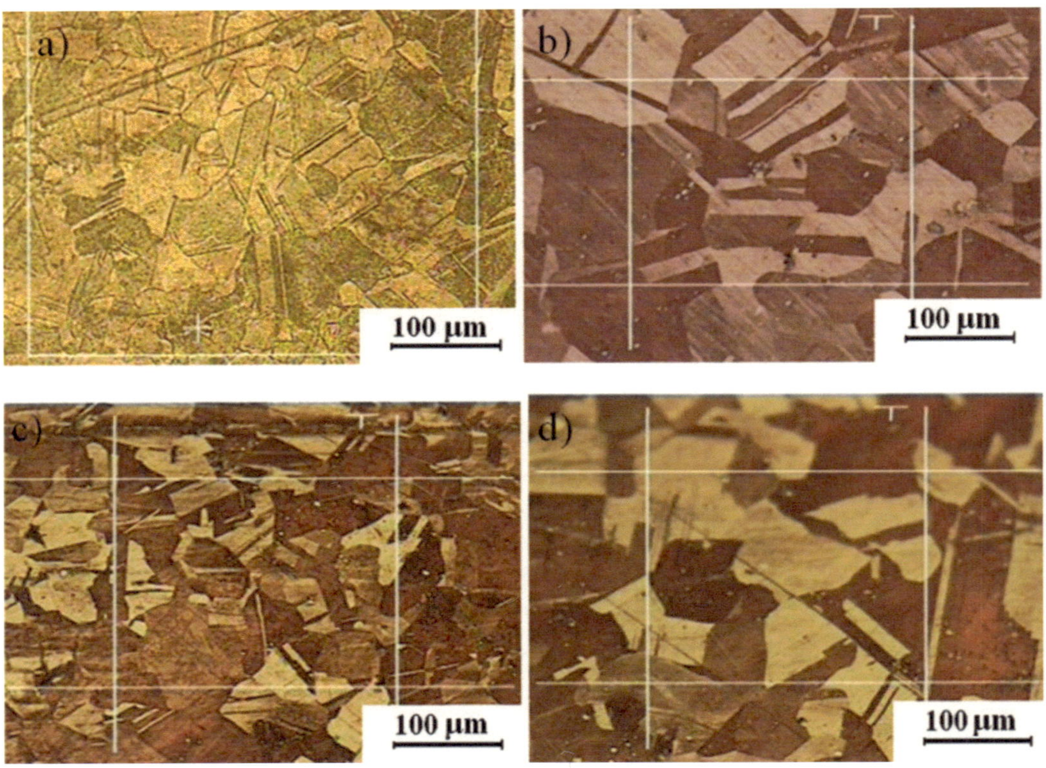

Figure 12. Optical photographs of the microstructure of the initial copper after pre-heat treatment: (a) initial microstructure, (b) annealing, (c) quenching, and (d) normalizing.

Figure 13. Optical photographs of the microstructure of the copper after three passes of "rolling-pressing" process: (a) initial microstructure, (b) annealing, (c) quenching, and (d) normalizing.

radial direction. Individual changes in the structure of copper after "rolling-pressing" were investigated using transmission electron microscopy (**Figure 14**).

During rolling after the first pass, there is a reduction in cross-border distances in both cross sections. Reducing of those distances is due to initial grains compression due to deformation. The creation of new borders occurs slowly, and all fragmentation occurs in the equal channel angular matrix. In accordance with the rule of Hall-Petch, the rolling in the first cycles leads to increasing the strength of parameters of copper by decreasing the interval between the boundaries in both cross sections.

Also, it was established that the second pass of combined process leads to the creation of the structure of mixed type: small recrystallized and deformed. Such structure is created due to the action of two processes: fragmentation in ECA-matrix and recrystallization during rolling. As the result, the copper with such structure has high ductility and strength. After the third pass, the copper structure shows a big part of large-angle boundaries.

TEM investigations have revealed that after two passes in the longitudinal direction elongated grains appear, which during next passes obtain more equiaxed grain form. Also there is an

Figure 14. Microstructure of copper after three passes of the process "rolling-pressing" in equal-channel step matrix, obtained with a transmission electron microscope: (a) initial microstructure, (b) annealing, (c) quenching, and (d) normalizing.

increase in the part of borders with large angles. After three passes, copper forms the structure with an equiaxed form and an average size of grains equals to 3.5 µm.

Complex investigation of different preliminary heat treatments on copper billets deformed by the "rolling-pressing" process revealed that preliminary heat treatment almost has no effect on the size of grains of pure copper.

5.2. Mechanical properties of copper before and after the "rolling-pressing" process

The tests for determining microhardness by Vickers were performed using the optical microscope Leica equipped with a set box for hardness testing (**Figure 15**).

The results of microhardness and average grain diameter of copper after thermal pretreatment are presented in **Table 2**. In contrast to other alloys, copper has the highest hardness after slow cooling in air, and after rapid cooling in water hardness becomes the lowest. After the "rolling-pressing" process, measurement of microhardness by Vickers, tests for tension and compression was conducted.

The strength characteristics are represented by values of yield stress and tensile stress; plastic characteristics are represented by the values of relative contraction and elongation (**Table 3**).

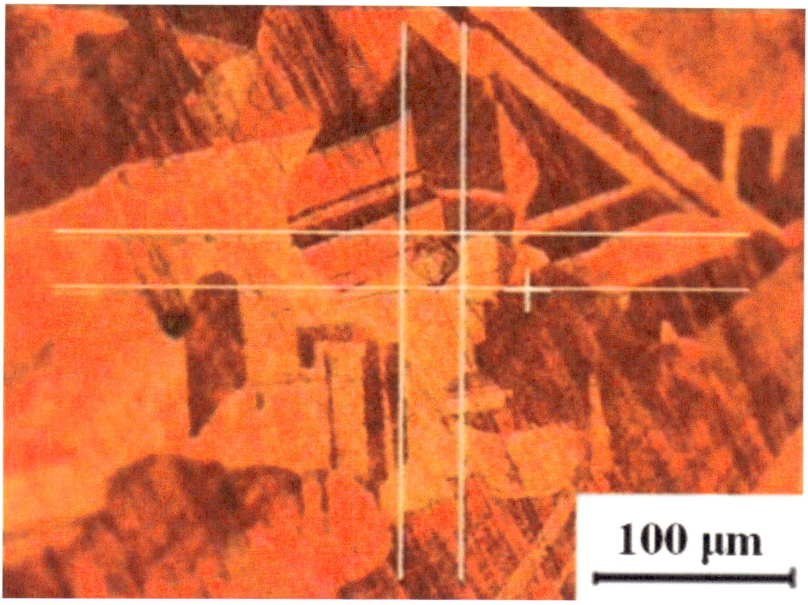

Figure 15. Determination of microhardness of copper.

Heat treatment process	Microhardness HV	Average grain diameter, μm
Initial microstructure	97.5	100
Annealing	104.6	125
Normalizing	109.5	105
Quenching	96.8	95

Table 2. Microhardness and average grain diameter of copper after preliminary heat treatment.

Heat treatment process	Microhardness HV	Yield stress, MPa	Tensile stress, MPa	ψ, %	δ, %
Initial microstructure	229.4	402	469	25.5	3.6
Annealing	215.7	360	374	36.3	8.0
Normalizing	236.8	402	502	22.2	5.7
Quenching	246.1	460	560	26.0	7.2

Table 3. Results of mechanical testing of copper after "rolling-pressing" process.

The most intensive hardening of copper occurs at relatively small degrees of deformation (two passes), then the process of hardening becomes slow, and during subsequent deformation, dynamic weakening occurs; as a result, the structure is less processed.

6. Conclusions

During the theoretical research of combined process "rolling-pressing" with equal-channel step matrix, the following results were obtained:

- Obtained empirical equations for determining the values of force of rolling, pressing, and back pressure from the matrix.
- Compiled a program for determining the optimal value of the angle of intersection of channels in the matrix, allowing you to quickly determine optimal conditions for the implementation of this combined process.
- Obtained formulas for determining diameters of the rolls, the values of which would provide the best capture angle of the workpiece.
- Studied, during computer simulation, parameters of the stress-strain state, temperature distribution, and the effect of the tilting of the workpiece on the process parameters. The results indicate the possibility of obtaining parts with sub-ultrafine-grained structure at realization of combined process "rolling-pressing."

After experiment for the study of influence of the "rolling-pressing" process on the microstructure of copper, the following results were established:

- After the first pass, the structure of copper is intensely reduced till about 40 μm.
- The creation of new borders occurs slowly; all fragmentation occurs in the ECA-matrix.
- After three passes in copper forms, the structure with an equiaxed form and average size of grains equal to 3.5 μm.
- A comprehensive study of the effect of different prethermal treatments on the structural changes in copper after the "rolling-pressing" process showed that preliminary heat treatment has little effect on the grain size of pure copper produced by severe plastic deformation.

At the same time, in almost identical grain size of the structure after the "rolling-pressing" process, the copper after quenching has greater values of the yield stress and tensile stress in comparison with copper obtained after other heat treatment processes.

Author details

Abdrakhman Naizabekov[1], Sergey Lezhnev[1]*, Evgeniy Panin[2] and Irina Volokitina[2]

*Address all correspondence to: sergey_legnev@mail.ru

1 Rudny Industrial Institute, Rudny, Kazakhstan

2 Karaganda State Industrial University, Temirtau, Kazakhstan

References

[1] Valiev RZ, Langdon TG. Principles of equal-channel angular pressing as a processing tool for grain refinement. Progress in Materials Science. 2006;**51**(7):881–981. DOI: 10.1016/j.pmatsci.2006.02.003

[2] Lezhnev SN, Naizabekov AB, Panin YA. Theoretical studies of the joint "extrusion-rolling" process aimed at making sub-ultra fine-grained structure metal. In: Tanger Ltd, editor. 20th Anniversary International Conference on Metallurgy and Materials; May 18-20. Brno, Czech Republic: Tanger Ltd; 2011. pp. 272–277

[3] Naizabekov A, Lezhnev S, Panin E, Koinov T. Theoretical grounds of the combined "rolling-equal-channel step pressing" process. Journal of Chemical Technology and Metallurgy. 2016;**51**(5):594–602.

[4] Naizabekov A, Lezhnev S, Panin E, Koinov T. Study of broadening in a combined process "rolling-pressing" using an equal-channel step die. Journal of Chemical Technology and Metallurgy. 2015;**50**(3):308–313.

[5] Fastykovsky AR, Peretyat'ko VN. Study of the effect of front back pressure on the length of the zones of sliding and sticking in the deformation zone during rolling. Izvestiya VUZ. Chernaya Metallurgiya. 2002;**2**:15–17.

[6] Naizabekov AB, Andreyashchenko VA. Determination of the effort during equal channel angular pressing. Technology of Production of Metals and Secondary Materials. 2006;**2**:120–124.

[7] Lezhnev S, Panin E. Investigation of the influence of geometric and technological factors on the stress-strain state of metal in the implementation of the combined rolling-pressing process. Advanced Materials Research. 2014;**936**:1918-1924. DOI: 10.4028/www.scientific.net/AMR.936.1918

[8] Naizabekov A, Lezhnev S, Panin E, Koinov T, Mazur I. Development and theoretical study of new scheme of realization of combined process "rolling-pressing" using equal channel step matrix. Journal of Chemical Technology and Metallurgy. 2017;**52**(2):186–192.

[9] Naizabekov AB, Lezhnev SN, Panin EA, Panin DA. Kinematics of the combined process "rolling-pressing" in equal channel step matrix. In: KSTU, editor. International Scientific-Practical Conference "Scientific-Technical Progress in Metallurgy"; October, 16-18. Temirtau, Kazakhstan: KSTU; 2007. pp. 238–242.

[10] Kopylov VI, Chuvildeev BN. The limit of grinding grains in equal-channel angular deformation. Metals. 2004;**1**:22–35.

[11] Lopatin YG. The evolution of the grain structure during deformation and annealing of microcrystalline materials obtained by equal channel angular pressing [thesis]. Moscow; 2010.

Chapter 8

Innovative Applications of Ultrafine-Grained Materials

Jie Xu, Bin Guo and Debin Shan

Additional information is available at the end of the chapter

http://dx.doi.org/10.5772/intechopen.69503

Abstract

This chapter focuses on multifunctional properties of ultrafine-grained (UFG) metallic materials processed by severe plastic deformation (SPD), such as enhanced mechanical properties, excellent superplasticity, and wear resistance. Based on these multifunctional properties, the potential innovative application for UFG materials processed by SPD is introduced in the next section, including innovative application in micro-forming, nano-implants, electro-connections, and sport engineering.

Keywords: ultrafine-grained material, properties, micro-forming, MEMS

1. Introduction

Materials experts have asserted that materials breakthroughs in the twentieth century required about 20 years from the time of invention to gain widespread market acceptance [1]. Ultrafine-grained (UFG) materials are used as a structural material due to these properties. Bulk nanostructured metallic materials also have been following this track. Twenty-five years ago, in 1988, there appeared a classic description of the application of severe plastic deformation (SPD) to bulk solids in order to achieve exceptional grain refinement to the submicrometer level [2]. Though a wide research started at the beginning of 1990, a great progress in commercial applications of UFG materials has been made just in the last few years. This chapter focused on multifunctional properties of ultrafine-grained metallic materials, including mechanical properties, superplasticity, wear resistance, etc. The innovative application of UFG materials was introduced in the following section, including in micro-forming and other commercial industries.

2. Multifunctional properties of ultrafine-grained materials

2.1. Enhanced mechanical properties

The grain size, d, plays a dominant role on the strength of polycrystalline metallic materials according to the Hall-Petch equation which states that the yield stress, σ_y, is given by [3, 4]

$$\sigma_y = \sigma_0 + k\,d^{-1/2} \tag{1}$$

where σ_0 is the friction stress and k is the Hall-Petch constant [1]. It follows from Eq. (1) that the strength increases with a decrease of grain size, and this leads to an ever-increasing interest in producing UFG materials with grain size of submicrometer or even nanometer level, which are processed by severe plastic deformation (SPD) techniques, including equal-channel angular pressing (ECAP) and high-pressure torsion (HPT). This means that UFG materials are anticipated to exhibit exceptional strength according to the Hall-Petch equation in Eq. (1).

For example, **Figure 1** shows the average Vickers microhardness values which were taken over the total surface of each disk for the AZ31 magnesium alloy processed by ECAP for up

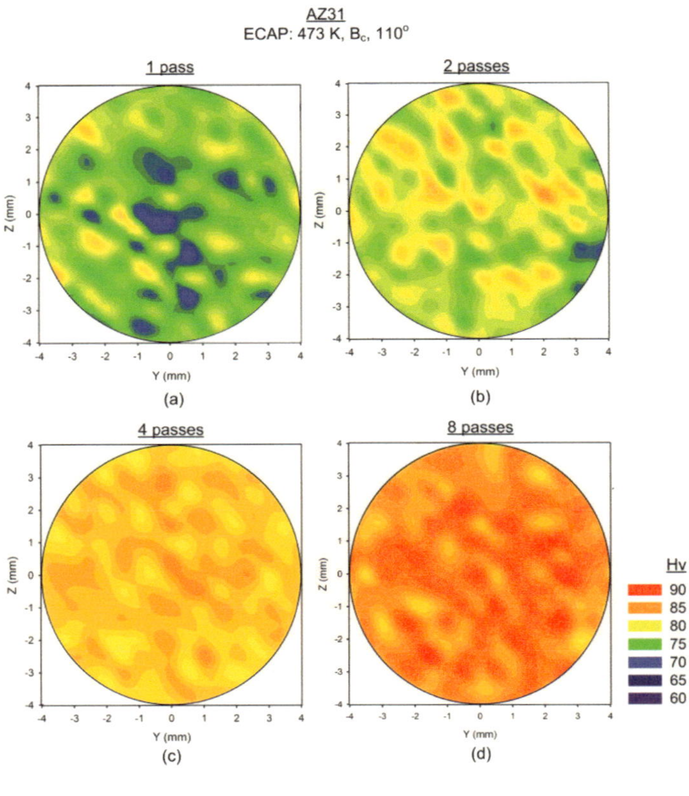

Figure 1. Distribution of microhardness of AZ31 processed by ECAP through (a) one pass, (b) two passes, (c) four passes, and (d) eight passes [5].

to eight passes [5]. The results demonstrate that the average value of the microhardness, Hv, increases significantly after one pass and continues to increase slowly up to eight passes. It is apparent in **Figure 1** that the samples processed by ECAP through one and two passes exhibit a higher average microhardness over the entire surface than the as-received sample as shown in **Figure 1(a)** and **(b)**. After four passes of ECAP, the sample achieves a reasonable level of homogeneity over the plane as shown in **Figure 1(c)**, and finally there is additional hardening and a general homogeneity after ECAP for eight passes in **Figure 1(d)**. These microhardness results demonstrate that the strength can be enhanced significantly by ECAP processing.

By comparison with the sample processed by ECAP, **Figure 2** shows the evolution in microhardness over one-quarter of the disk for the same AZ31 alloy processed by HPT through (a) 1/4, (b) 1, (c) 5, and (d) 10 turns [6]. There is a gradual evolution toward higher microhardness values with increasing numbers of turns after HPT processing. There is a reasonable level of hardness homogeneity which is achieved across the HPT disks through ten turns with a saturation hardness value of Hv ≈ 125 as shown in **Figure 2** [6]. This gradual development of hardness homogeneity is consistent with several experimental reports for magnesium alloys processed by HPT [7–10]. There are many other papers reporting the enhanced strength of UFG materials processed by SPD methods [11–15].

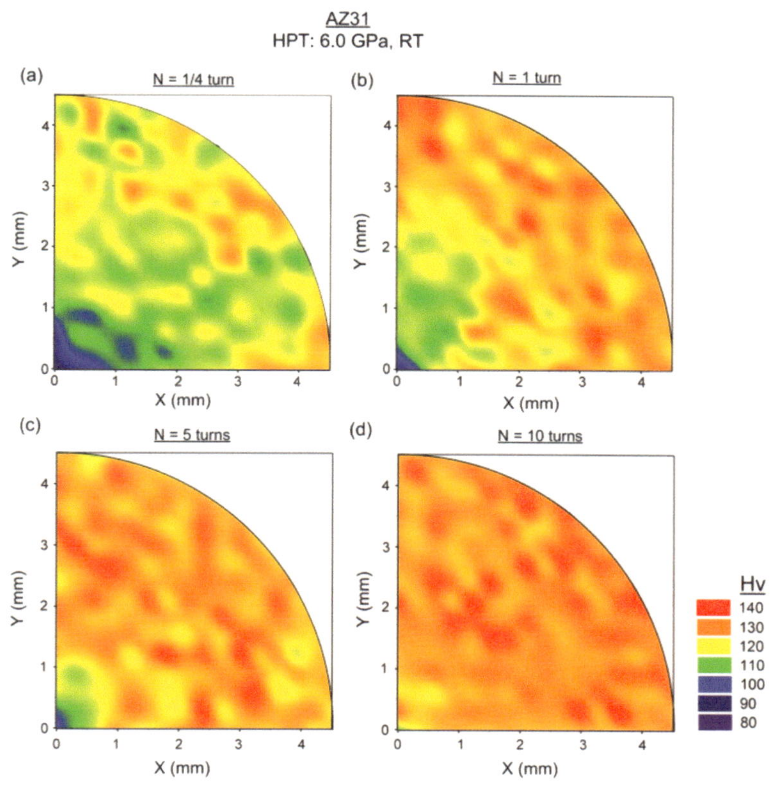

Figure 2. The variation of microhardness over one-quarter area of the disks processed by HPT through (a) 1/4, (b) 1, (c) 5, and (d) 10 turns[6].

High ductility in metallic materials is another very important property, which is essential for metal-forming operations as well as to avoid catastrophic failure in load-bearing applications during their service life. However, most of the UFG materials processed by SPD demonstrate significantly higher strength than the coarse-grained (CG) counterparts but have a relatively low ductility. Various strategies to improve low ductility of the UFG materials have been proposed, which can be divided into two groups of "mechanical" strategies and "microstructural" strategies [16, 17]. For example, there are different finds of strength and ductility in high-purity Cu with initial coarse grains, cold rolling (CR) with reduction ratio of 60%, and after ECAP processing up to 16 passes. The results for the ECAP-processed Cu demonstrate an enhanced strength with good ductility similar to the CG sample [12]. As shown in **Figure 3**, for Cu and Al, CR (the reduction in thickness is marked by each datum point) increases the yield strength but decreases the elongation to failure [18, 19]. The extraordinary combination of high strength and high ductility shown in **Figure 3** for the nanostructured Cu and Ti after SPD processing clearly sets them apart from the other CG metals [20].

2.2. Excellent superplasticity

Superplasticity is a well-recognized mechanical property in polycrystalline metallic materials that have the ability to pull out to a very high elongation without any significant necking in tension. The superplastic flow mechanism is dominated by the process of grain boundary sliding (GBS) in which the small grains slide over each other in response to the applied stress [21]. The GBS needs intragranular slip as an accommodation mechanism, and this slip

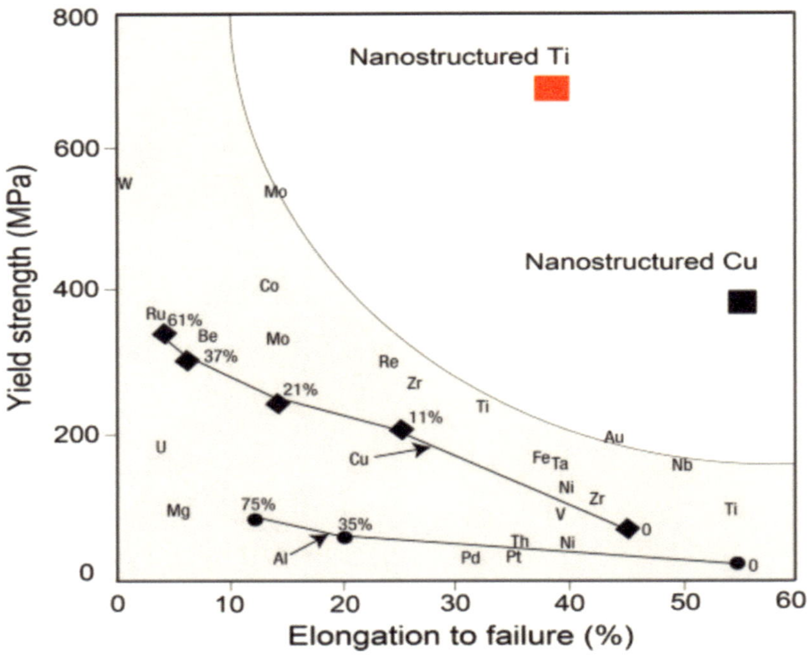

Figure 3. The extraordinary combination of high strength and high ductility in metals processed by SPD [20].

is held up at subgrain boundaries in CG materials. Accordingly, the UFG materials processed by SPD have an opportunity to achieving good superplasticity due to submicrometer grain size with high fraction of high-angle boundaries [22–32].

For example, there is no superplasticity after cold rolling (CR) because of the presence of low-angle sub-boundaries. By comparison, the presence of UFG structures can lead to exceptional superplasticity at elevated temperatures. An excellent superplasticity in an UFG Al-3% Mg-0.2% Sc alloy after processing by ECAP through eight passes was found, and the maximum elongation of 2280% can be obtained after tension tests at strain rate of 3.3×10^{-2} s^{-1} and the temperature of 673 K [33]. The highest elongation of 3030% was recorded after tension at a strain rate of 1.0×10^{-4} s^{-1} and the temperature of 473 K in a ZK60 Mg-5.5% Zn-0.5% Zr alloy processed by extrusion and ECAP as shown in **Figure 4** [34]. This is the highest elongation in a Mg alloy processed under any condition and one of the highest elongations recorded in any materials processed by ECAP [35]. In addition, the absence of any visible necking within the gauge length in **Figure 4** demonstrates conclusively that this is a true superplastic flow [36].

Processing by HPT can also produce UFG materials that have a potential for achieving superplastic elongation in tension. An example is shown in **Figure 5** for the superplasticity in

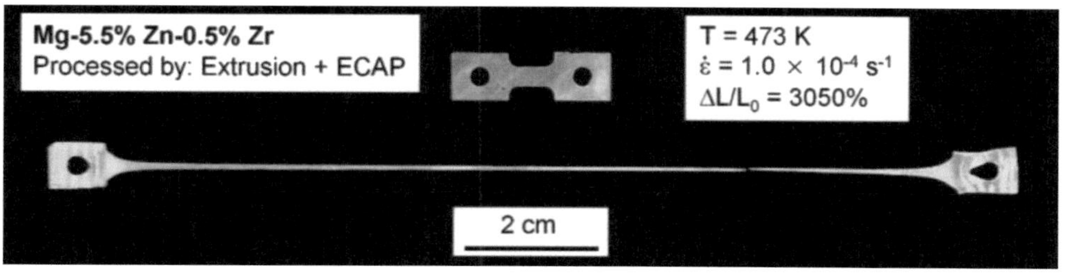

Figure 4. Exceptional superplasticity in a ZK60 alloy processed by ECAP [34].

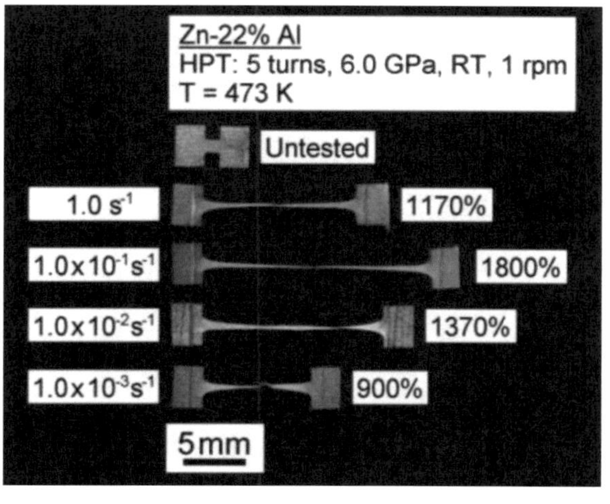

Figure 5. Superplasticity in the Zn-22% Al alloy after processing by HPT [37].

tension at 473 K using different strain rates in a Zn (22%)-Al alloy processed by HPT through five turns at room temperature under an applied pressure of 6.0 GPa [37]. It is evident from **Figure 5** that very high elongations may be achieved at strain rates in the vicinity of 10^{-1} s^{-1}, whereas there is a clear evidence for necking in the two samples pulled at the slowest strain rates. A tabulation of superplastic data for samples prepared by HPT shows that the elongation of 1800% visible in **Figure 5** is the highest elongation reported to date for any material processed by HPT [37]. However, the highest elongation in the same UFG Zn-22% Al alloy processed by ECAP occurs at 10^{-2} s^{-1} [38]. High strain rate superplasticity can be achieved by using ECAP or HPT. Compared with the sample processed by ECAP, the optimum superplasticity for the sample processed by HPT correctly occurs at faster strain rate, but maximum elongation is reduced. The elongation is reduced because HPT samples have very small gauge sections.

In the recent report, there is an instructive comparison of the superplasticity in various materials processed by ECAP and HPT with other processing techniques as shown in **Figure 6** [39], where the superimposed on each diagram are the appropriate ranges for UFG Al alloy processed by ECAP and HPT indicated by the dashed ovals fill in the diagrams first developed 20 years ago [40]. It is readily apparent from **Figure 6** that both the ECAP and HPT processing methods extend the plastic forming rate of the given materials with faster strain rates and higher elongation. The expanded ranges generally overlap with the ranges associated with powder metallurgy (PM) materials. This expansion in the range of strain rate reveals an important advantage for the SPD processing without any contamination and/or porosity

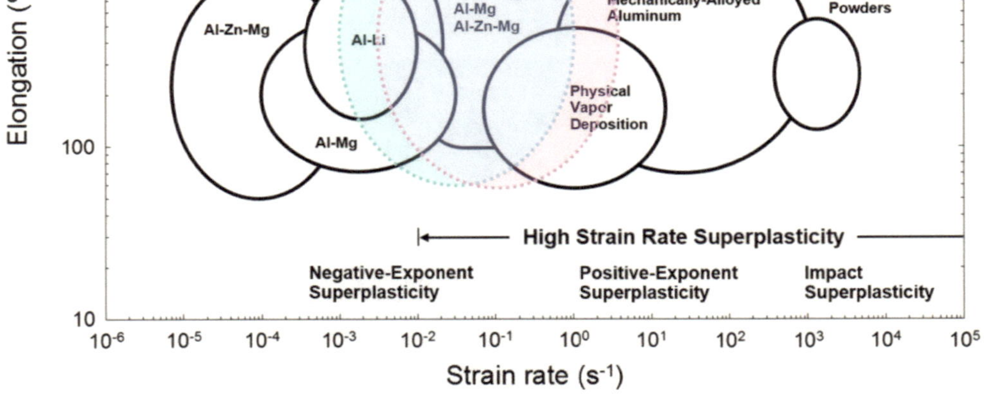

Figure 6. A plot of elongation versus strain rate for a series of Al alloys produced using different processing methods [39].

in using PM methods. Thus, processing of SPD techniques demonstrates a very important approach for extending the future applications of numerous simple metals and alloys [39].

2.3. Wear resistance

Wear resistance is an important property for UFG materials in order to evaluate their potential for use as structural components [41]. The wear of sliding surfaces can occur by one or more wear mechanisms, including adhesion, abrasion, fatigue wear, corrosive wear, and fretting. For the metallic materials, the wear volume under abrasive and some adhesive wear models is generally assumed to be inversely proportional to the hardness of the materials according to the traditional Archard relationship which is given by [42]

$$V = K\frac{LN}{H} \tag{2}$$

where V is the wear loss of the volume, N is the applied force, L is the sliding distance, K is the wear coefficient, and H is the hardness on wear surface of the material. Because the UFG materials processed by SPD techniques normally have much higher hardness values than the conventional CG materials, it is critical to have superior wear resistance for UFG materials. However, there has been a disagreement in this regard among researchers [41].

There are a number of studies reporting an improved wear resistance in UFG materials produced by ECAP and HPT. For example, the dry sliding wear tests of an aluminum alloy processed by ECAP method showed that the wear mass loss decreased significantly with increasing of the numbers of ECAP passes [43]. A similar enhanced wear resistance property was also presented in an Al-Mg-Si alloy processed by ECAP [44]. An investigation of friction and wear behavior revealed that grain size was the important factor determining the transition from elasto-hydrodynamic lubrication to the boundary lubrication regions [45, 46]. An investigation of the aluminum bronze alloy processed by ECAP demonstrated that the coefficient of friction decreased with increasing numbers of ECAP passes and accordingly the wear resistance was improved significantly after ECAP processing [47, 48]. Similarly, a characterization of the dry sliding wear behaviors of Cu-0.1 wt.% Zr alloy and AZ31 alloy processed by ECAP was investigated [49, 50], and the wear volume loss of the samples processed by ECAP becomes much lower than the annealed alloy as shown in **Figure 7** due to the higher microhardness introduced by ECAP processing [49]. Processing by ECAP can produce bulk materials with significantly enhanced mechanical properties due to the grain refinement, and therefore the wear loss of the ECAP-processed alloy is much smaller than for the annealed alloy. Some papers are now available on the wear behavior of commercial purity Ti processed by HPT method. Compared with the CG pure Ti, the wear resistance of pure Ti processed by HPT was improved significantly both in dry and wet sliding tests [51, 52].

On the other hand, there are also some other contradictory results on wear property in UFG materials. For example, the wear resistance of some UFG materials processed by ECAP was lower than for the as-received CG materials [53]. For example, the dry sliding wear tests of an Al-1050 alloy were conducted with the as-received condition and UFG materials with grain size of ~1.3 μm after ECAP processing through eight passes [54]. The UFG samples

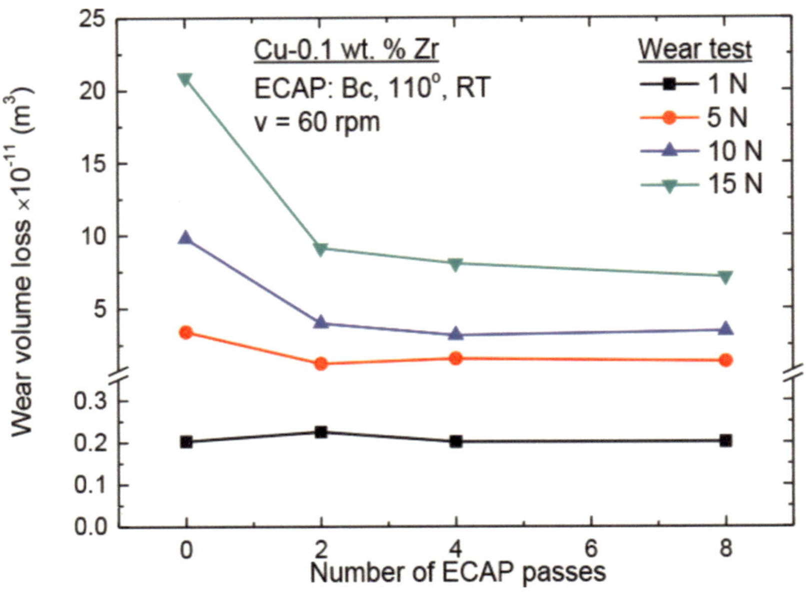

Figure 7. The wear volume loss versus the number of ECAP passes for a sliding time of 600 s under normal loads of 1, 5, 10, and 15 N.

have a similar coefficient of friction (COF) and the higher wear loss than the as-received sample although the microhardness value is improved significantly after ECAP processing. An investigation of UFG AISI 1024 steel processed by a warm multiaxial forging technique showed that there is no obvious improvement on wear resistance property though the strength property can be enhanced significantly due to the effects of higher density of grain boundaries and submicrometer-sized cementite particles [55]. There is a surprising result that there is no corresponding improvement in the wear resistance in pure titanium processed due to the occurrence of oxidative wear with an abrasive effect [56]. As a consequence of these varying reports, it is readily apparent that further investigations should be further conducted in order to evaluate the wear behavior of UFG materials processed by SPD techniques.

3. Innovative application of ultrafine-grained materials

It is well established that SPD techniques are very effective in producing UFG materials with submicrometer or even nanoscale grain sizes, and these materials have superior mechanical properties including high strength and, if the fine grains are reasonably stable, a good superplastic capability at elevated temperatures [57, 58]. Despite a wide research on SPD techniques started at the beginning of 1990, very significant progress in the commercialization of UFG materials has been made in the recent years. In this section, the innovative application of UFG metallic materials processed by SPD is discussed.

http://dx.doi.org/10.5772/intechopen.69503

3.1. Potential application in micro-forming technology

Micro-forming is defined as the production of parts or structures having at least two dimensions in the submillimeter range, which becomes an attractive option in the manufacturing of these products because of its advantages for mass production with controlled forming quality, high production rate, and low cost [59–61]. Nevertheless, although the knowledge of tool design and fabrication techniques are now well developed for the conventional macro-forming, there is an evidence that the occurrence of size effects may lead to a breakdown in these basic plastic deformation theory when the specimen dimensions are scaled down to the micro/mesoscale [62, 63]. In practice, if there are only a few grains in the micro-parts, the response to the applied forces will show significant variations, and the reproducibility of the mechanical properties will become a serious problem in any micro-forming processes [64]. Hopefully, there is a way to solve grain size effects in micro/mesoforming by applying UFG materials with submicrometer or even nanoscale grain sizes produced by SPD techniques [65–68] because ultrafine grains can improve the micro-formability, surface roughness, and good mechanical properties of the MEMS components [69–71].

However, micro-deformation behavior changes from dislocation dominated in large grains to grain boundary dominated in small-grain regimes when the grain size decreases to the submicron range. For example, the deformation behavior in UFG pure aluminum processed by ECAP and post-annealed specimens at room temperature (RT) was investigated, and the results show that different work hardening behaviors were observed during macro-compression test when the grain size increased from 0.35 to 45 μm [72]. The strain rate also has an obvious effect on micro-compression behavior of UFG pure aluminum, and the results demonstrate that a lower strain rate causes activation of micro-shear banding [73], and the deformation mechanism may be related to grain boundary sliding in UFG pure aluminum [74]. Thus, it is believed that grain boundary sliding and grain rotations are the main deformation mechanism in the UFG materials processed by SPD techniques. However, there is only limited information available on micro-forming when the material grain size is reduced to the submicrometer even nanoscale level although these problems and limitations are beginning to attract attentions within the materials science community. At the present time, Le et al. [75] investigated the influence of grain size ranging from 0.5 to 5.2 μm on the deformation behavior by compression tests at macroscale in aluminum prepared by a spark plasma sintering method. The results indicate that there is a strong correlation between deformation microstructure and grain size in the micrometer regime. Okamoto et al. [76] investigated the specimen and grain size effects on micro-compression formation behavior in the electrodeposited nanocrystalline copper with average grain size of 360, 100, and 34 nm. The results show that the deformation mechanisms with nanocrystalline grains are different from those for pillar with submicron grain size from the surface microstructure of deformed micropillars. There is a significant micro-deformation difference between the CG and UFG materials. For example, the micro-deformation behavior is transferred from work hardening to slight strain weakening with decreasing of grain size during micro-compression. The microstructural evolution results show that a lot of low-angle grain boundaries and recrystallized fine grains are formed inside of the original large grains in CG pure aluminum. By contrast, ultrafine grains are kept in UFG pure aluminum, which are similar to the original microstructure before micro-compression. Meanwhile, there is an obvious transition from nonuniform deformation to uniform deformation after micro-compression testing with decreasing of grain size as shown in **Figure 8a–e**. The nonuniform deformation can be improved significantly, and the compressed specimens using UFG pure Al are cylindrical with a smooth surface as shown in **Figure 8d** and **e** [77, 78]. Research on the micro-extrusion of UFG aluminum showed that the material flow became more uniform because more grains were deformed during micro-extrusion [79]. Similarly, an investigation of the effect of specimen size on tensile testing with UFG and CG pure copper demonstrates that the uniform elongation increases with increasing specimen thickness and decreasing gauge length. In addition, the failure mode changed gradually from shear to normal tensile failure with increasing of specimen thickness [80, 81]. Therefore, the surface roughness and coordinated deformation ability can be significantly improved during micro-compression with UFG materials, which demonstrates that they have a potential application in micro-forming at ambient temperature.

The ductility at micro/mesoscale is another method to evaluate the UFG materials whether the alloy has the potential for use in micro-forming applications. The mechanical properties confirm the general behavior anticipated for UFG metals including a strengthening at ambient temperature through the Hall-Petch relationship and a decrease in yield stress and higher ductilities when testing at elevated temperatures. For example, an investigation of the superplastic micro-forming of the magnesium AZ91 alloy with a UFG microstructure showed that the grain size and the transition from superplastic flow to non-superplastic flow were the main parameters controlling the micro-formability [82]. A UFG AZ31 magnesium alloy with average grain size of ~110 nm processed by HPT for 10 turns under an imposed pressure of 6.0 GPa shows a highest elongation of ~400% testing at a temperature of 423 K and a strain rate of 1.0×10^{-4} s^{-1} [6]. This elongation of the sample processed by HPT is more than two times larger than the elongation of ~192% recorded in the same alloy processed by ECAP through eight passes testing at 472 K [5]. Thus, the micro-tensile testing of the UFG AZ31 magnesium

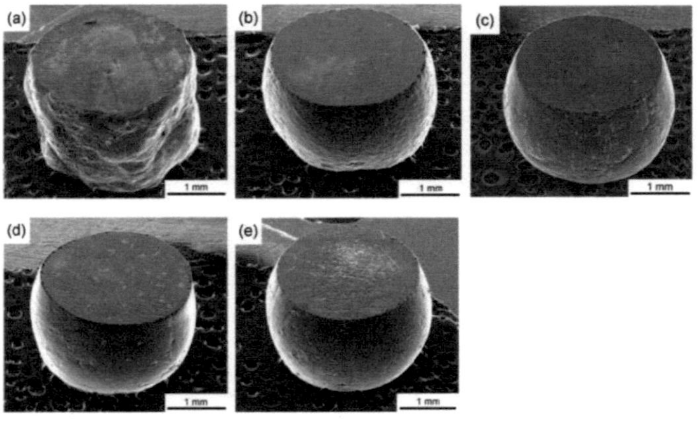

Figure 8. Surface topographies of the compressed sample with grain size of (a) ~150, (b) ~25, (c) ~4, (d) ~1.5, and (e) ~1.3 μm when compressed with fixed specimen diameter 2 mm [77].

alloy processed by HPT suggests the possibility of obtaining a true superplastic property at a testing temperature which is much lower than for the same samples processed by ECAP.

To evaluate the micro-formability of UFG materials, a micro-V-groove die with a width of 100 μm and the V angle of 60° was proposed as shown in **Figure 9** [83]. The micro-coining tests were conducted with the as-drawn and UFG AZ31 magnesium alloy at the temperatures ranging from 298 to 523 K. After micro-coining tests, the surface shape of the embossed specimen was measured, and the filling area Af was calculated from the measurement data. The filling ratio Rf of the filling area Af to the V-groove area Av was used to evaluate the formability after micro-embossing. **Figure 9** shows the different filling behaviors during micro-coining with the as-drawn AZ31 magnesium alloy and UFG AZ31 alloy processed by HPT. For the as-drawn AZ31 magnesium alloy, the percentage of flowed area, Rf, increases slowly with increasing temperature from room temperature to 423 K and then increases abruptly up to 523 K. In contrast, the filling ratio, Rf, increases significantly when the embossing temperature in UFG AZ31 is elevated from 298 to 423 K and then continues to increase slowly with increasing of embossing temperature up to 523 [84]. Thus, UFG AZ31 alloy processed by HPT exhibits an excellent micro-formability by superplastic deformation, which is expected to become one of most useful materials to fabricate MEMS components with complicated structures.

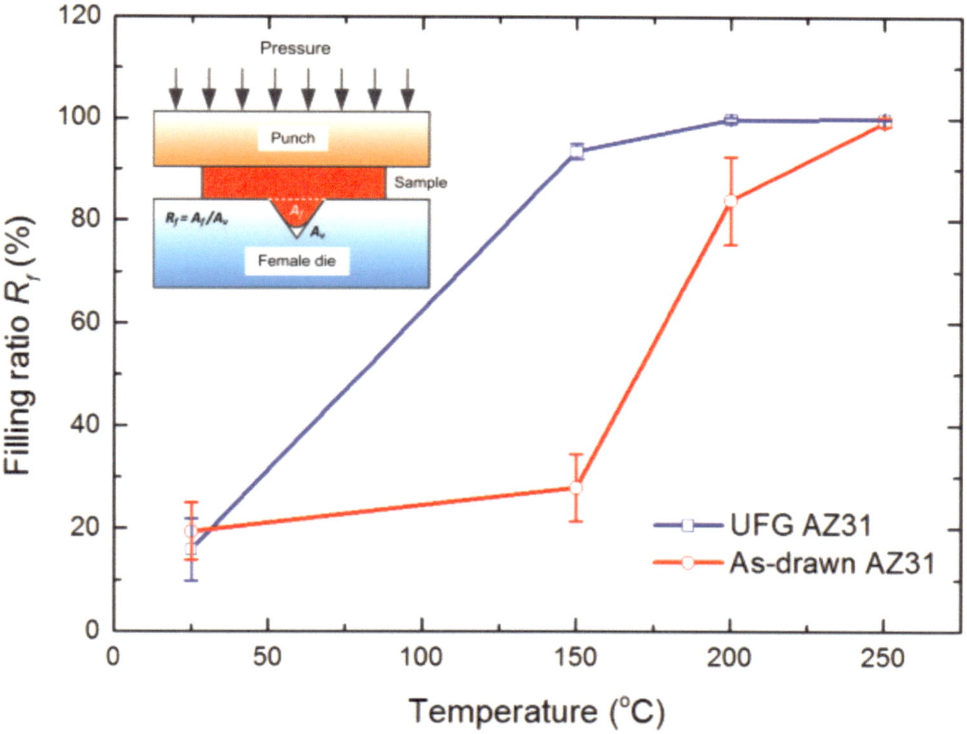

Figure 9. Plots of filling ratio versus the increasing of embossing temperature for UFG AZ31 alloy processed by HPT and as-drawn AZ31 alloy [83].

Based on these results, it is concluded that the SPD-processed UFG materials have a strong potential for use in micro-forming applications at elevated temperature. For example, a UFG pure Al with average grain size of ~1.0 μm produced by ECAP was adopted for micro-hot-embossing processes using a novel micro-embossing tool that was designed with a self-adaptive adjustment and a vacuum mounting system [84]. The microarray channels are fabricated with feather widths from 5 to 100 μm at the temperature of 523 K under a force of 4 kN followed by a dwell time of 600 s as shown in **Figure 10**. The embossed micro-channels of 100 μm in width are clearly formed with a good geometrical transferability and no obvious defects as shown in **Figure 10a**. The straight side walls are replicated from the micro-silicon dies, but the top surface becomes rough and even with the decreasing of channel widths, as shown in **Figure 10b–d**. These results demonstrate that the filling quality is mainly attributed to the channel dimension compared to the grain size at the given micro-embossing conditions [84].

Figure 11 shows the comparison of the profile measurements for the micro-channels that are 25 μm in width were embossed under the same experimental conditions using CG pure Al and UFG pure Al after ECAP processing through eight passes [84]. The filling problem of CG pure Al with an average grain size of ~300 μm is much more serious for micro-embossing at 25 μm in width because there are some wrinkles and uneven channels after micro-embossing. During the micro-embossing tests of the CG pure Al, the micro-channel on the silicon die is filled by a single grain deformation in the transverse direction because the grain size of CG pure Al is much larger than the channel width. So the material flow behavior is different at the grain boundary and at the edge of the micro-channels, which leads to an inclined surface and wrinkles. By contrast, the micro-embossing of UFG pure Al at the same temperature produces smooth micro-channels, and the patterns on the silicon mold are fully transferred to the UFG pure Al plate. These results demonstrate that the UFG pure Al has much better formability than the CG pure Al. Therefore, micro-hot-embossing of UFG pure Al has good potential for application

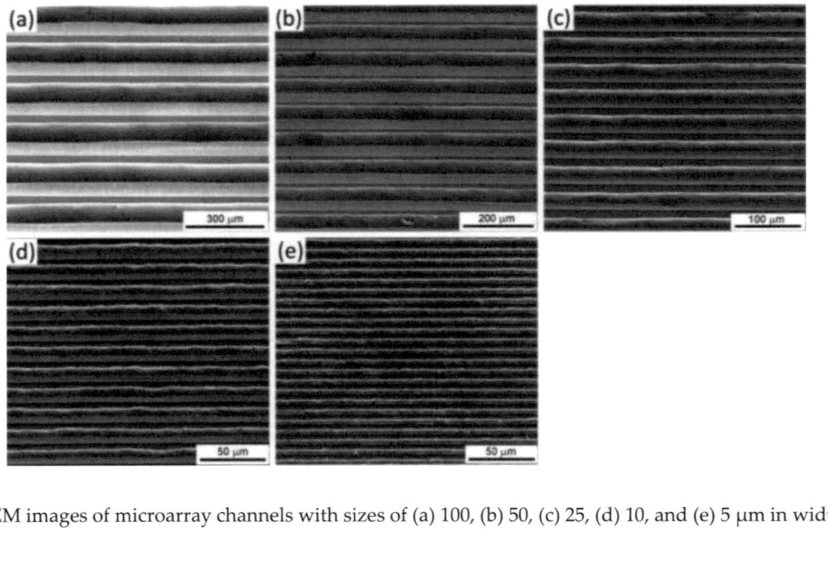

Figure 10. SEM images of microarray channels with sizes of (a) 100, (b) 50, (c) 25, (d) 10, and (e) 5 μm in width [84].

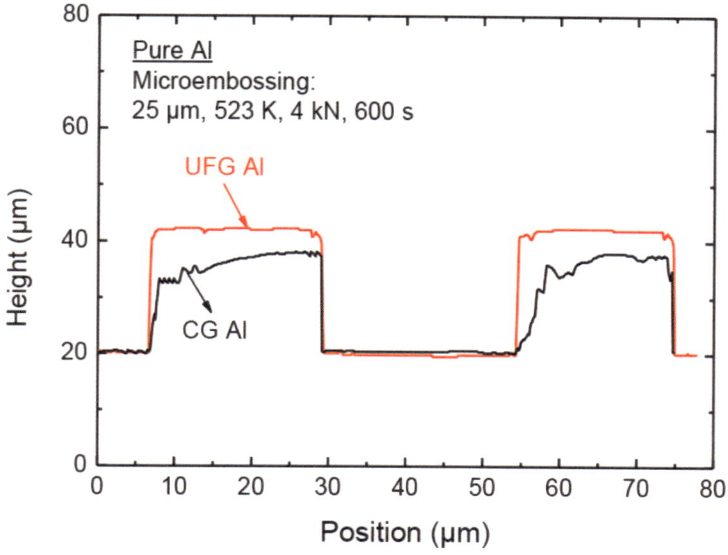

Figure 11. Comparison of the filling quality using UFG and CG pure Al [84].

in the fabrication of micro-parts with the micro-forming mold equipped with self-adaptive adjustment and a vacuum mounting system [84].

The UFG materials processed by SPD appear to provide a significant potential for use in micro-forming applications at elevated temperatures due to their enhanced mechanical properties at the room temperature and improved ductility at the elevated temperatures. However, the present investigation demonstrates that there is also an excellent micro-formability when using UFG pure aluminum at ambient temperature. The micro-tensile testing shows that the UFG pure Al processed by ECAP has excellent mechanical properties compared with the CG pure Al. The highest elongation of ~72% after ECAP processing suggests a good potential for using this material in micro-forming process at ambient temperature [85]. Moreover, micro-compression testing shows that the UFG pure Al produced by ECAP has improved the deformation compatibility by comparison with the CG sample and benefits to filling quality during micro-forming. This was confirmed by successfully using micro-forming to fabricate a micro-turbine from UFG pure aluminum at ambient temperature as shown in **Figure 12**. The perfection of this micro-turbine is a direct consequence of the high forming quality and the generally uniform mechanical properties of this material. The high strength and high level of homogeneity are also confirmed directly by microhardness measurements. These results demonstrate that there is an excellent potential for using UFG materials to fabricate micro-parts with high accuracy, high strength, and a high level of uniformity [85].

3.2. Commercial applications of ultrafine-grained materials

Application and commercialization of UFG materials are associated with three primary points: their superior properties, their efficient fabrication, and the possibility to produce

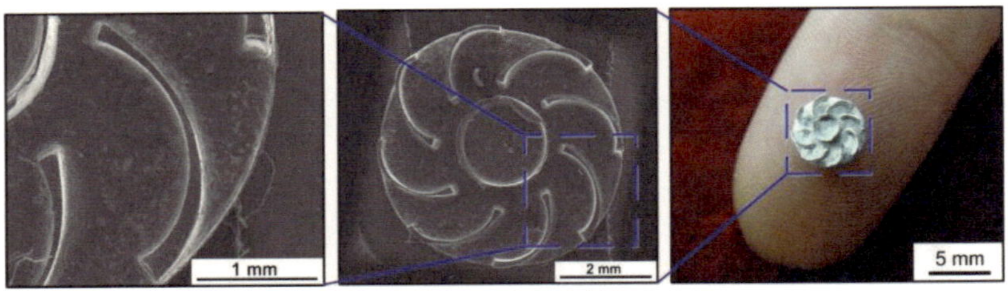

Figure 12. Micro-turbine of UFG pure aluminum formed at ambient temperature [85].

cutting-edge products from these materials [86]. Below are the examples of UFG materials processed by SPD for their commercial applications in biomedical engineering, electrical engineering, and sports.

The UFG pure titanium processed by ECAP-Conform from the Ufa State Aviation Technical University under the management of professor Valiev has been used as trademark application to manufacture dental implants in the company "Timplant" (Ostrava, Czech Republic) since 2006 [87]. The UFG Ti with ultimate strength of 1350 MPa enabled design of thin dental implant with diameter of 2.0 mm, which serves as fully functional pillar, and it can be inserted into very thin bones. Another advantage of smaller dental implants is less damage induced into jawbone during surgery intervention [88]. To date, these dental implants have been certified according to the European standard EN ISO 13485:2003. **Figure 13a** illustrates the Nanoimplant®, which is installed into the body of an 18-year-old patient with thin jawbones between teeth 11 and 13. Another implant with the diameter of 2.4 mm was inserted to the right-side position 12 as shown in **Figure 13b** and **c**. Two nanoimplants with two temporary crowns made in the same day as implants were inserted in the patient left the dental office. After 6 weeks, the final metal-ceramic crowns were fixed on the implants [89]. One of the possible next dental implant products with UFG Ti produced by SPD was manufactured and sold by basic implant systems under the trademark Biotanium in the USA beginning in 2011 [86]. Thus, the small-diameter dental implants made from UFG Ti are possible to replace standard ones made from Ti-4Al-6V alloy, since the UFG pure Ti is characterized not only by the improved mechanical strength and fatigue life but also by better biocompatibility compared to the conventional Ti-4Al-6V alloy.

UFG pure copper, aluminum, and aluminum alloy would be an innovative solution for electro-connections in high-voltage current converter due to the improved mechanical property without reduction of their electrical conductivity or even with its significant improvement. For example, very thin Cu and Al-2% Fe wires with a final diameter of 0.08 mm were successfully drawn from the ring sample processed by HPT for N = 1 revolution as shown in **Figure 14** [90]. A 25:1 area reduction after wire drawn can be achieved from the HPT processed samples, but the wire draw from the as-cast state failed after 12:1 reduction [90]. The electrical conductivity of the wires ranges from 49–51 IACS% and increased to 52–54% after aging at 473 K for 1 h. These results demonstrate that there is a large potential to further improve the electrical conductivity with an optimized aging treatment [91].

http://dx.doi.org/10.5772/intechopen.69503

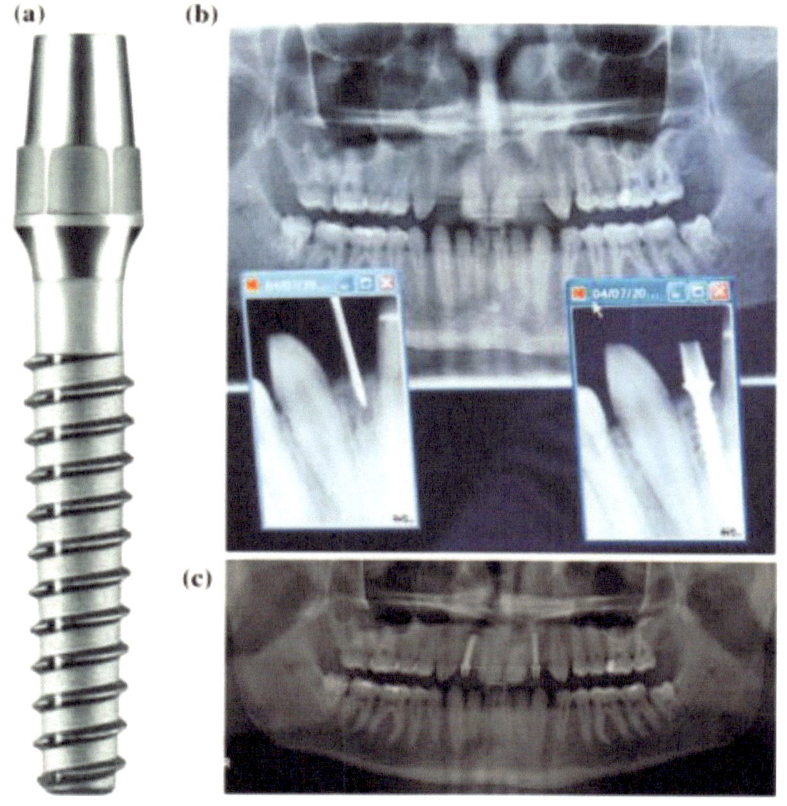

Figure 13. (a) Dental implant from nanostructured Ti and (b and c) X-ray photographs after surgery and control photograph after incorporation of dental implants into human jaw [89].

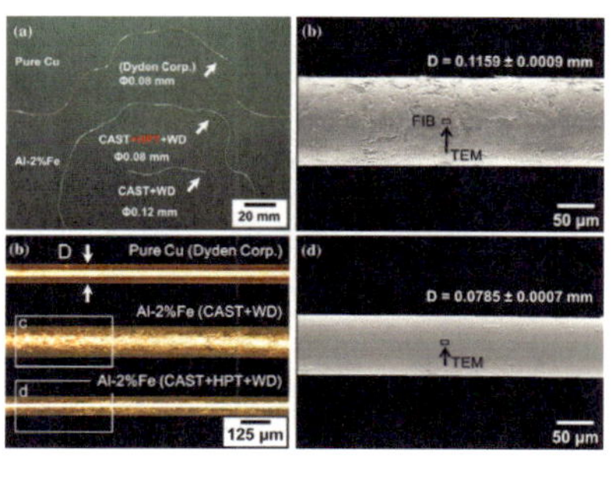

Figure 14. (a) Photograph and (b) optical micrograph of pure Cu and Al-2 % Fe as-drawn wires. SEM images showing surface condition of wires drawn from c as cast state and d HPT-processed state [90].

Producers of sport devices/equipment can also benefit from the UFG metals, particularly where high strength and low weight are required. The UFG materials could find applications in high-performance golf, bicycles, tennis, hockey, mountain equipment, etc. One of

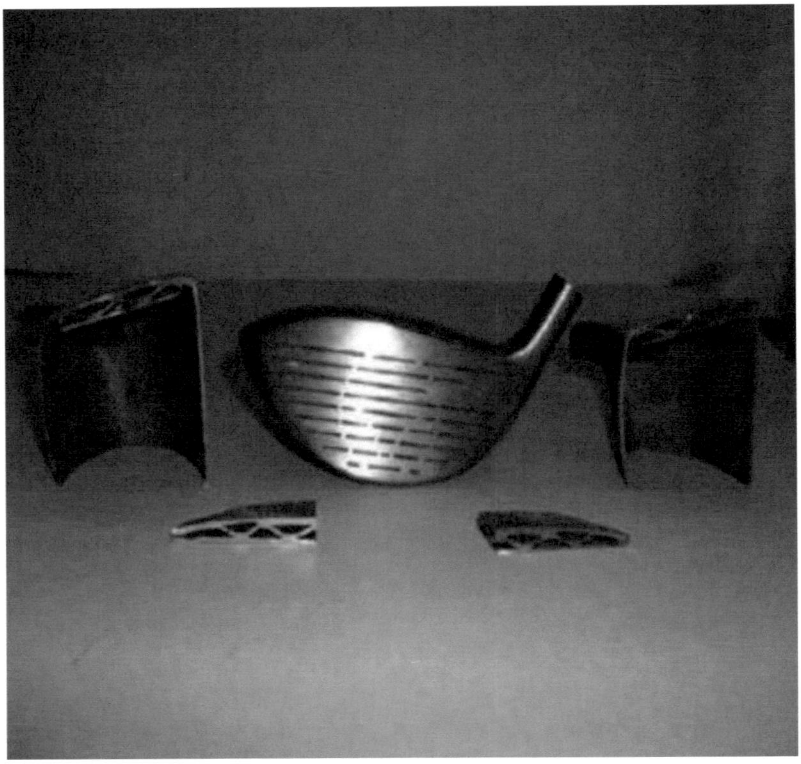

Figure 15. Components of golf club made from nanostructured Ti-6Al-4V alloy [93].

the important examples is nano-dynamic high-performance golf balls, which have a hollow nanostructured titanium core. The core material is manufactured using the UFG chip from Purdue University [92]. The Institute for Metals Superplasticity Problems (Russia) has developed a technology for the fabrication of golf club components from UFG Ti-6Al-4V alloy with grain size of 200 nm as shown in **Figure 15** [93]. The method for producing the goffer-type face using UFG or nanostructured metals and inserts provided processing faces characterized by enhanced strength and high-impact efficiency. This technology allowed a reduction in weight of a golf club along with increase of ball's flight distance due to increased restitution factor [93]. These application results demonstrate wide commercial potentialities for applying UFG materials processed by SPD.

Author details

Jie Xu*, Bin Guo and Debin Shan

*Address all correspondence to: xjhit@hit.edu.cn

Harbin Institute of Technology, Harbin, China

References

[1] Eagar T. Bringing new materials to market. Technology Review. 1995;98:42-49

[2] Langdon TG. Twenty-five years of ultrafine-grained materials: Achieving exceptional properties through grain refinement. Acta Materialia. 2013;61:7035-7059

[3] E.O. Hall. The deformation and ageing of mild steel: III discussion of results. Proc. Phys. Soc. Lond. B 1951;**64**:747-752

[4] N.J. Petch. The cleavage strengh of polycrystals. J. Iron. Steel. Inst. 1953;**174**:25-28

[5] Xu J, Shirooyed M, Wongsa-Ngam J, Shan D, Guo B, Langdon TG. Hardness homogeneity and micro-tensile behavior in a magnesium AZ31 alloy processed by equal-channel angular pressing. Materials Science and Engineering: A. 2013;**586**:108-114

[6] Xu J, Wang X, Shirooyed M, Xing G, Shan D, Guo B, Langdon TG. Microhardness, microstructure and tensile behavior of an AZ31 magnesium alloy processed by high-pressure torsion. Journal of Materials Science. 2015;**50**:7424-7436

[7] Vrátná J, Janeček M, Čížek J, Lee DJ, Yoon EY, Kim HS. Mechanical properties and microstructure evolution in ultrafine grained AZ31 alloy processed by severe plastic deformation. Journal of Materials Science. 2013;**45**:4705-4712

[8] Kawasaki M, Figueiredo RB, Huang Y, Langdon TG. Interpretation of hardness evolution in metals processed by high-pressure torsion. Journal of Materials Science. 2014;**49**:6586-6596

[9] Malheiros LRC, Figueiredo RB, Langdon TG. Grain size and microhardness evolution during annealing of a magnesium alloy processed by high-pressure torsion. Journal of Materials Research and Technology. 2015;**4**:14-17

[10] Vrátná J, Janeček M, Gubicza J, Krajňák T, Yoon EY, Kim HS. Evolution of microstructure and hardness in AZ31 alloy processed by high pressure torsion. Materials Science and Engineering: A. 2015;**625**:98-106

[11] Furukawa M, Horita Z, Nemoto M, Valiev RZ, Langdon TG. Factors influencing the flow and hardness of materials with ultrafine grain sizes. Philosophical Magazine A. 1998;**78**:203-216

[12] Valiev RZ, Alexandrov IV, Zhu YT, Lowe TC. Paradox of strength and ductility in metals processed by severe plastic deformation. Journal of Materials Research. 2002;**17**:5-8

[13] Valiev RZ, Sergueeva AV, Mukherjee AK. The effect of annealing on tensile deformation behavior of nanostructured SPD titanium. Scripta Materialia. 2003;**49**:669-674

[14] Horita Z, Ohashi K, Fujita T, Kaneko K, Langdon TG. Achieving high strength and high ductility in precipitation-hardened alloys. Advanced Materials 2005;**17**:1599-1602

[15] Sabirov I, Murashkin MYu, Valiev RZ. Nanostructured aluminium alloys produced by severe plastic deformation: new horizons in development. Materials Science and Engineering: A. 2013;**560**:1-24

[16] Zhao Y, Zhu YT, Lavernia EJ. Strategies for improving tensile ductility of bulk nanostructured materials. Advanced Engineering Materials. 2010;**12**:769-778

[17] Ma E. Eight routes to improve the tensile ductility of bulk nanostructured metals and alloys. JOM. 2006;**58**(4):49-53

[18] Parker ER. Materials Data Book for Engineers and Scientists. New York: McGraw-Hill; 1967

[19] Brandes EA, Brook GB. In: Smithells Metals Reference Book, 7th ed. Oxford: Butterworth–Heinemann; 1992 (Chap. 22.)

[20] Valiev RZ. Nanostructuring of metals by severe plastic deformation for advanced properties. Nature Materials. 2004;**3**:511-516

[21] Langdon TG. An evaluation of the strain contributed by grain boundary sliding in superplasticity. Materials Science and Engineering: A. 1994;**174**:225-230

[22] Lee S, Utsunomiya A, Akamatsu H, Neishi K, Furukawa M, Horita Z, Langdon TG. Influence of scandium and zirconium on grain stability and superplastic ductilities in ultrafine-grained Al–Mg alloys. Acta Materialia. 2005;**50**:553-564

[23] Nikulin I, Kaibyshev R, Sakai T. Superplasticity in a 7055 aluminum alloy processed by ECAE and subsequent isothermal rolling. Materials Science and Engineering: A. 2005;**407**:62-70

[24] Kaibyshev R, Shipilova K, Musin F, Motohashi Y. Achieving high strain rate superplasticity in an Al–Li–Mg alloy through equal channel angular extrusion. Materials Science and Technology. 2005;**21**:408-418

[25] Turba K, Ma´lek P, Cieslar M. Superplasticity in an Al–Mg–Zr–Sc alloy produced by equal-channel angular pressing. Materials Science and Engineering: A. 2007;**462**:91-94

[26] Mishra RS, Valiev RZ, McFadden SX, Islamgaliev RK, Mukherjee AK. High-strain-rate superplasticity from nanocrystalline Al alloy 1420 at low temperatures. Philosophical Magazine A. 2001;**81**:37-48

[27] Sakai G, Horita Z, Langdon TG. Grain refinement and superplasticity in an aluminum alloy processed by high-pressure torsion. Materials Science and Engineering: A. 2005;**393**:344-351

[28] Dobatkin SV, Bastarache EN, Sakai G, Fujita T, Horita Z, Langdon TG. Grain refinement and superplastic flow in an aluminum alloy processed by high-pressure torsion. Materials Science and Engineering: A. 2005;**408**:141-146

[29] Xu C, Dobatkin SV, Horita Z, Langdon TG. Superplastic flow in a nanostructured aluminum alloy produced using high pressure torsion. Materials Science and Engineering: A. 2009;**500**:170-175

[30] Sabbaghianrad S, Kawasaki M, Langdon TG. Microstructural evolution and the mechanical properties of an aluminum alloy processed by high-pressure torsion. Journal of Materials Science. 2012;**47**:7789-7795

[31] Kawasaki M, Foissey J, Langdon TG. Development of hardness homogeneity and superplastic behavior in an aluminum-copper eutectic alloy processed by high-pressure torsion. Materials Science and Engineering: A. 2013;**561**:118-125

[32] Alhamidi A, Horita Z. Grain refinement and high strain rate superplasticity in aluminium 2024 alloy processed by high pressure torsion. Materials Science and Engineering: A. 2015;**622**:139-145

[33] Komura S, Horita Z, Furukawa M, Nemoto M, Langdon TG. An evaluation of the flow behavior during high strain rate superplasticity in an Al-Mg-Sc alloy. Metallurgical and Materials Transactions A. 2001;**32**:707-716

[34] Figueiredo RB, Langdon TG. Record superplastic ductility in a magnesium alloy processed by equal-channel angular pressing. Advanced Engineering Materials. 2008;10:37-40

[35] Figueiredo RB, Langdon TG. Fabricating ultrafine-grained materials through the application of severe plastic deformation: a review of developments in Brazil. Journal of Materials Research and Technology. 2012;**1**:55-62

[36] Langdon TG. Fracture processes in superplastic flow. Metal Science. 1982;**16**:175-183

[37] Kawasaki M, Langdon TG. Developing superplasticity and a deformation mechanism map for the Zn-Al eutectoid alloy processed by high-pressure torsion. Materials Science and Engineering: A. 2011;**528**:6140-6145

[38] Kawasaki M, Langdon TG. Grain boundary sliding in a superplastic zinc-aluminum alloy processed using severe plastic deformation. Materials Transactions. 2008;**49**:84-89

[39] Kawasaki M, Langdon TG. Review: Achieving superplastic properties in ultrafine-grained materials at high temperatures. Journal of Materials Science. 2016;**51**:19-32

[40] Kawasaki M, Langdon TG. Principles of superplasticity in ultrafine-grained materials. Journal of Materials Science. 2007;**42**:1782-1796

[41] Gao N, Wang CT, Wood RJK, Langdon TG. Tribological properties of ultrafine-grained materials processed by severe plastic deformation. Journal of Materials Science. 2012;**47**:4779-4797

[42] Archard JF. Contact and rubbing of flat surfaces, Journal of Applied Physics. 1953;**24**: 981-988

[43] Li J, Wongsa-Ngam J, Xu J, Shan D, Guo B, Langdon TG. Wear resistance of an ultrafine-grained Cu-Zr alloy processed by equal-channel angular pressing. Wear. 2015;**326-327**:10-19

[44] Xu J, Wang X, Zhu X, Shirooyeh M, Wongsa-Ngam J, Shan DB, Guo B, Langdon TG. Dry sliding wear of an AZ31 magnesium alloy processed by equal-channel angular pressing, Journal of Materials Science. 2013;**48**:4117-4127

[45] Abd El Aal MI, EI Mahallaw N, Shehata FA, Abd EI Hameed M, Yoon EY, Kim HS. Wear properties of ECAP-processed ultrafine grained Al-Cu alloys. Materials Science and Engineering: A. 2010;**527**:3726-3732

[46] Ortiz-Cuellar E, Hernandez-Rodriguez MAL, García-Sanchez E. Evaluation of the tribological properties of an Al-Mg-Si alloy processed by severe plastic deformation. Wear. 2011;**271**:1828-1832

[47] Moshkovich A, Perfilyev V, Lapsker I, Gorni D, Rapoport L. The effect of grain size on stribeck curve and microstructure of copper under friction in the steady friction state. Tribology Letters. 2011;**42**:89-98

[48] Moshkovich A, Perfilyev V, Gorni D, Lapsker I, Rapoport L. The effect of Cu grain size on transition from EHL to BL regime (Stribeck curve). Wear. 2011;**271**:1726-1732

[49] Gao LL, Cheng XH. Microstructure and dry sliding wear behavior of Cu-10%Al-4%Fe alloy produced by equal channel angular extrusion. Wear. 2008;**265**:986-991

[50] Gao LL, Cheng XH. Effect of ECAE on microstructure and tribological properties of Cu-10%Al-4%Fe alloy. Tribology Letters. 2007;**27**:221-225

[51] Faghihi S, Li D, Szpunar JA. Tribocorrosion behaviour of nanostructured titanium substrates processed by high-pressure torsion. Nanotechnology. 2010;**21**:485703

[52] Wang CT, Gao N, Gee MG, Wood RJK, Langdon TG. Effect of grain size on the micro-tribological behavior of pure titanium processed by high-pressure torsion. Wear. 2012;**280-281**:28-35

[53] Kim YS, Lee T, Park KT, Kim WJ, Shin DH. Dry sliding wear behavior of commercial purity aluminum and low carbon steel by severe plastic deformation techniques. in: Zhu YT, Langdon TG, Mishra RS, Semiatin SL, Saran MJ, Lowe TC, editors. Ultrafine Grained Materials II. USA: John Wiley & Sons, Inc.; 2013. pp. 409-418

[54] Wang CT, Gao N, Wood RJK, Langdon TG. Wear behavior of an aluminum alloy processed by equal-channel angular pressing. Journal of Materials Science. 2011;**46**:123-130

[55] Padap AK, Chaudhari GP, Nath SK. Mechanical and dry sliding wear behavior of ultrafine-grained AISI 1024 steel processed using multiaxial forging. Journal of Materials Science. 2010;**45**:4837-4845

[56] Purcek G, Saray O, Kul O, Karaman I, Yapici GG, Haouaoui M, Maier HJ. Mechanical and wear properties of ultrafine-grained pure Ti produced by multi-pass equal-channel angular extrusion. Materials Science and Engineering: A. 2009;**517**:97-104

[57] Valiev RZ, Islamgaliev RK, Alexandrov IV. Bulk nanostructured materials from severe plastic deformation. Progress in Materials Science. 2000;**45**:103-189

[58] Valiev RZ, Estrin Y, Horita Z, Langdon TG, Zehetbauer MJ, Zhu YT. Producing bulk ultrafine-grained materials by severe plastic deformation. JOM. 2006;**58**(4):33-39

[59] Geiger M, Kleiner M, Eckstein R, Tiesler N, Engel U. Microforming. CIRP Annals—Manufacturing Technology. 2001;**50**:445-462

[60] Engel U, Eckstein R. Microforming—From basic research to its realization. Journal of Materials Processing Technology. 2002;**35**:125-126

[61] Vollertsen F, Niehoff HS, Hu Z. State of the art in micro forming. International Journal of Machine Tools and Manufacture. 2006;**46**:1172-1179

[62] Vollertsen F, Biermann D, Hansen HN, Jawahir IS, Kuzman K. Size effects in manufacturing of metallic components. CIRP Annals—Manufacturing Technology. 2009;**58**:566-587

[63] Fu MW, Chan WL. A review on the state-of-the-art microforming technologies. International Journal of Advanced Manufacturing Technology. 2013;**67**:2411-2437

[64] Janssen PJM, De Keijser TH, Geers MGD. An experimental assessment of grain size effects in the uniaxial straining of thin Al sheet with a few grains across the thickness. Materials Science and Engineering: A. 2006;**419**:238-248

[65] Estrin Y, Janecek M, Raab GI, Valiev RZ, Zi A. Severe plastic Deformation as a means of producing ultra-fine-grained net-shaped micro electro-mechanical systems parts. Metallurgical and Materials Transactions A. 2007;**38**:1906-1909

[66] Zhilyaev AP, Langdon TG. Using high-pressure torsion for metal processing: Fundamentals and applications. Progress in Materials Science. 2008;**53**:893-979

[67] Kwan CCF, Wang Z. Cyclic deformation of ultra-fine grained commercial purity aluminum processed by accumulative roll-bonding. Materials. 2013;**6**:3469-3481

[68] Kim WJ, Sa YK. Micro-extrusion of ECAP processed magnesium alloy for production of high strength magnesium micro-gears. Scripta Materialia. 2006;**54**:1391-1395

[69] Ma X, Lapovok R, Gu C, Molotnikov A, Estrin Y, Pereloma EV, Davies CHJ, Hodgson PD. Deep drawing behaviour of ultrafine grained copper: modelling and experiment. Journal of Materials Science. 2009;**44**:3807-3812

[70] Xu J, Zhu X, Shi L, Shan D, Guo B, Langdon TG. Micro-forming using ultrafine-grained aluminum processed by equal-channel angular pressing. Advanced Engineering Materials. 2015;**636**:3527-360

[71] Xu J, Shi L, Wang C, Shan D, Guo B. Micro hot embossing of micro-array channels in ultrafine-grained pure aluminum using a silicon die. Journal of Materials Processing Technology. 2015;**225**:375-384

[72] Yu CY, Sun PL, Kao PW, Chang CP. Mechanical properties of submicron-grained aluminum. Scripta Materialia. 2005;**52**:359-363

[73] Sabirov I, Barnett MR, Estrin Y, Hodgson PD. The effect of strain rate on the deformation mechanisms and the strain rate sensitivity of an ultra-fine-grained Al alloy. Scripta Materialia. 2009;**61**:181-184

[74] Wang M, Shan A. Effect of strain rate on the tensile behavior of ultra-fine grained pure aluminum. Journal of Alloys and Compounds. 2008;**455**:L10-L14

[75] Le GM, Godfrey A, Hansen N, Liu W, Winther G, Huang X. Influence of grain size in the near-micrometre regime on the deformation microstructure in aluminium. Acta Materialia. 2013;**61**:7072-7086

[76] Okamoto NL, Kashioka D, Hirato T, Inui H. Specimen- and grain-size dependence of compression deformation behavior in nanocrystalline copper. International Journal of Plasticity. 2014;**56**:173-183

[77] Xu J, Zhu X, Shan D, Guo B, Langdon TG. Effect of grain size and specimen dimensions on micro-forming of high purity aluminum. Materials Science and Engineering: A. 2015;**646**:207-217

[78] Xu J, Li J, Zhu X, Fan G, Shan D, Guo B. Microstructural evolution at micro/meso-scale in an ultrafine-grained pure aluminum processed by equal-channel angular pressing with subsequent annealing treatment. Materials. 2015;**8**:7447-7460

[79] Rosochowski A, Presz W, Olejnik L, Richert M. Micro-extrusion of ultra-fine grained aluminium. International Journal of Advanced Manufacturing Technology. 2007;**33**:137-146

[80] Zhao YH, Guo YZ, Wei Q, Dangelewicz AM, Xu C, Zhu YT, Langdon TG, Zhou YZ, Lavernia EJ. Influence of specimen dimensions on the tensile behavior of ultrafine-grained Cu. Scripta Materialia. 2008;**59**:627-630

[81] Zhao YH, Guo YZ, Wei Q, Topping TD, Dangelewicz AM, Zhu YT, Langdon TG, Lavernia EJ. Influence of specimen dimensions and strain measurement methods on tensile stress-strain curves. Materials Science and Engineering: A. 2009;**525**:68-77

[82] Kim WJ, Yoo SJ, Kim HK. Superplastic microforming of Mg-9Al-1Zn alloy with ultra-fine-grained microstructure. Scripta Materialia. 2008;**59**:599-602

[83] Xu J, Xing X, Shan D, Guo B, Langdon TG. An evaluation of formability using micro-embossing on an ultrafine-grained magnesium AZ31 alloy processed by high-pressure torsion. MATEC Web of Conference. 2015;**21**:09005

[84] Xu J, Shi L, Wang C, Shan D, Guo B. Micro hot embossing of micro-array channels in ultrafine-grained pure aluminum using a silicon die. Journal of Materials Processing Technology. 2015;**225**:375-384

[85] Xu J, Zhu X, Shi L, Shan D, Guo B, Langdon TG. Advanced Engineering Materials. 2015;**17**:1022-1033

[86] I. Sabirow, N.A. Enikeev, M. Yu. Murashkin, R.Z. Valiev. Bulk Nanostructured Materials with Multifunctional Properties. Springer; 2015. Switzerland, 125 p

[87] www.timplant.cz

[88] Valiev RZ, Semenova IP, Latysh VL, Rack H, Lowe TC, Petruzelka J, Dluhos L, Hrusak D, Sochova J. Nanostructured titanium for biomedical applications. Advanced Engineering Materials. 2008;**10**:B15-B17

[89] Mishnaevsky Jr L, Levashov E, Valiev RZ, Segurado J, Sabirov I, Enikeev N, Prokoshkin S, Solov'yov AV, Korotitsky A, Gutmanas E, Gotman I, Rabkin E, Psakh'e S, Dluhos L, Seefeldt M, Smolin A. Nanostructured titanium-based materials for medical implants: modeling and development. Materials Science and Engineering: R. 2014;**81**:1-19

[90] Cubero-Sesin JM, In H, Arita M, Iwaoka H, Horita Z. High-pressure torsion for fabrication of high-strength and high-electrical conductivity Al micro-wires. Journal of Materials Science. 2014;**49**:6550-6556

[91] Champion Y, Brechet Y. Effect of grain size reduction and geometrical confinement in fine grained copper: Potential applications as a material for reversible electrical contacts. Advanced Engineering Materials. 2010;**12**:798-802

[92] Azushima A, Kopp R, Korhonen A, Yang DY, Micari F, Lahoti GD, Groche P, Yanagimoto J, Tsuji N, Rosochowski A, Yanagida A. Severe plastic deformation (SPD) processes for metals. CIRP Annals—Manufacturing Technology. 2008;**57**:716-735

[93] Safiullin AR, Safiullin RV, Kruglov AA. Application of nanostructured Ti alloys for producing a face for a gold club. Reviews on Advanced Materials Science. 2010;**25**:281-285